"十三五"职业教育国家规划教材

普通高等职业教育计算机系列规划教材

云操作系统应用
（OpenStack）

李 腾 武春岭 主 编

路 亚 马巾凌 于奎伟 副主编

电子工业出版社

Publishing House of Electronics Industry

北京·BEIJING

内 容 简 介

本书讲述开栈（OpenStack）云计算、整体框架、安装部署、代码剖析及扩展开发等内容。以 OpenStack（Mitaka）为蓝本，使用 VMware 模拟实际的物理平台，利用案例搭建和学习 OpenStack 手工和自动化安装、配置和维护云计算环境，详细介绍了 OpenStack 的部署和运行，对 OpenStack 配置文件进行修改定制，强调实践操作，达到熟练运用 OpenStack 系统的目的。

本书适合高职高专和应用型本科学生学习，也可作为从事"云计算"领域工作相关技术人员的参考用书。

未经许可，不得以任何方式复制或抄袭本书之部分或全部内容。
版权所有，侵权必究。

图书在版编目（CIP）数据

云操作系统应用：OpenStack/李腾，武春岭主编. —北京：电子工业出版社，2017.9
普通高等职业教育计算机系列规划教材
ISBN 978-7-121-32314-0

Ⅰ.①云… Ⅱ.①李… ②武… Ⅲ.①计算机网络－高等职业教育－教材 Ⅳ.①TP393

中国版本图书馆 CIP 数据核字（2017）第 181828 号

策划编辑：徐建军（xujj@phei.com.cn）
责任编辑：郝黎明　　特约编辑：张燕虹
印　　刷：涿州市般润文化传播有限公司
装　　订：涿州市般润文化传播有限公司
出版发行：电子工业出版社
　　　　　北京市海淀区万寿路 173 信箱　邮编 100036
开　　本：787×1 092　1/16　印张：11.25　字数：288 千字
版　　次：2017 年 9 月第 1 版
印　　次：2023 年 2 月第 14 次印刷
定　　价：29.00 元

凡所购买电子工业出版社图书有缺损问题，请向购买书店调换。若书店售缺，请与本社发行部联系，联系及邮购电话：（010）88254888，88258888。
质量投诉请发邮件至 zlts@phei.com.cn，盗版侵权举报请发邮件至 dbqq@phei.com.cn。
本书咨询联系方式：（010）88254570。

序 言
Introduction

  云计算自 2006 年由 Google 首席执行官 Eric Schmidt 正式提出，发展至今已逾 10 年。随着互联网时代计算日趋网络化、泛在化和智能化，面对高性能、大数据、高可靠的信息处理需求，云计算基于分布式处理、网络存储、虚拟化、负载均衡等技术，按需、易扩展的 IT 资源交付与服务模式已在金融、气象、电子商务、政务、医疗、企业管理领域被广泛采用。目前，亚马逊、微软、谷歌、百度、阿里巴巴、腾讯等网络运营商均提供自己的公有云服务。云计算作为国家"互联网+"战略的核心基础，必将呈现出巨大的产业发展活力和人才需求。

  国务院于 2015 年 1 月发布的《关于促进云计算创新发展培育信息产业新业态的意见》指出：鼓励普通高校、职业院校、科研院所与企业联合培养云计算相关人才，加强学校教育与产业发展的有效衔接，为云计算产业发展提供高水平智力支持。2015 年 10 月，教育部将"云计算技术与应用"专业列入高职专业目录。截至 2017 年，全国有 109 家高职院校开始招收"云计算技术与应用"专业学生。

  "云计算技术与应用"专业课程体系涵盖目前国内云计算行业技能型岗位人才所需的基本知识与技能。本书通过借鉴云计算行业企业前沿技术与项目开发实践，基于开源 OpenStack 架构，着重于 Linux Shell、OpenStack 云计算基础平台技术、虚拟化技术、云计算网络技术、云存储技术、Web 应用开发、Android 移动应用开发、Hadoop 大数据平台与应用开发、数据中心运维等核心知识的介绍与实战技能训练。该课程设计遵循"任务驱动、项目导向"原则，突出"技术应用能力、工程实践能力与职业竞争力"的培养。满足高职高专"云计算技术与应用"专业技能人才培养目标的要求。

  本书主要由重庆电子工程职业学院教师和中国电子科技集团公司南京第五十五研究所专家共同编写。不仅可作为高职院校云计算相关专业的课程教材，也可作为云计算行业岗位人才培训教材或参考资料。我们相信，随着云计算技术日渐成熟，与人工智能、大数据、VR 技术的融合发展，应用领域进一步拓展，云计算产业规模将不断扩大，对各类人才尤其是技能型应用人才的需求将持续增长。本系列教材的出版必将为"云计算技术与应用"专业建设和人才培养起到积极的推动作用。

中国通信工业协会信息安全与云计算校企联盟
全国云计算大数据职教集团

前 言
Preface

随着物联网、互联网的迅速发展，网络上流动的海量数据时刻需要处理，而传统的技术已无法满足当前的需要。云计算作为新一轮的信息技术革命，使得大量的应用运行在云端，许多企业、高校和政府部门也会根据实际需求建立自己的私有云。这些私有云可以在企业内部根据不同的部门、不同的业务或不同的租户来定制和分配所需的资源。虚拟化是云计算的底层技术和核心内容，能够有效地整合资源、降低能耗，并充分提高硬件的利用率，此外还能简化管理，提高数据中心的容灾能力。由于这些显著的优势，越来越多的企业使用虚拟化技术来搭建自己的私有云平台。在众多的虚拟化产品中，OpenStack"开源、开放、免费"的特点深深吸引着众多企业，仅需投入很少的费用就能建设一套低成本、不受厂商技术绑定、不侵犯知识产权的虚拟化或私有云平台，对于众多企业充满着无法抵御的诱惑。

本书重点介绍认证服务 Keystone、镜像服务 Glance、计算服务 Nova、网络部署服务 Neutron、对象存储服务 Swift、块存储服务 Cinder、编排服务 Heat、云网络和云主机的创建。对于学习者理解和搭建 OpenStack 或整个云计算体系有很大的帮助。第 1 章介绍云计算概念、发展历史、云计算体系架构和云计算平台 OpenStack，以及分享经典云计算解决案例；第 2 章介绍虚拟化技术和分类，以及 KVM 的安装、使用和虚拟机管理；第 3 章介绍 Openstack 环境的准备和配置；第 4 章介绍认证服务 Keystone 的概念及相关服务的安装配置；第 5 章介绍镜像服务 Glance 的相关服务的安装配置和镜像的制作；第 6 章介绍计算服务 Nova 的架构及原理和各个节点的相关配置；第 7 章介绍网络部署服务 Neutron 的网络基础知识及各节点的相关配置；第 8 章介绍对象存储服务 Swift 的概念及各节点配置，以及创建账户 Ring、容器 Ring 和创建对象 Ring；第 9 章介绍安装和配置 Web 服务 Dashboard；第 10 章介绍块存储服务 Cinder 的基本概念及各节点相关配置；第 11 章介绍编排服务 Heat 的基本概念、数据库配置、创建服务凭证和 API 端点与配置 Heat；第 12 章介绍云网络和云主机的创建；第 13 章介绍 OpenStack 典型架构实例。

为了使读者在学习时能直观地了解每个步骤的结果，本书对每个命令执行完的界面（窗口）都进行了完整的展示，故对展示的界面（窗口）没有按章排序编号和给出图题。

本书由重庆电子工程职业学院的李腾、武春岭担任主编，重庆电子工程职业学院路亚、马巾凌和华云数据技术开发有限公司于奎伟任副主编。南京第五十五所技术开发有限公司工程师参与了本书的案例设计和案例测试，在此表示衷心的感谢。重庆电子工程职业学院的卢兴俊、陈易、蒋鹏、黄建明等学生在本书的编写过程中一直参与案例测试和文字校对工作，在此也一并表示感谢。

为了方便教师教学，本书配有电子教学课件，请有此需要的教师登录华信教育资源网（www.hxedu.com.cn）注册后免费进行下载，如有问题可在网站留言板留言或与电子工业出版社联系（E-mail：hxedu@phei.com.cn）。

虽然我们精心组织，努力工作，但错误之处在所难免；同时由于编者水平有限，书中也存在诸多不足之处，恳请广大读者给予批评和指正，以便在今后的修订中不断改进。

编 者

目录 Contents

第1章 云计算概述 ·· (1)
 1.1 云计算简介 ·· (1)
  1.1.1 云计算概念与特征 ·· (1)
  1.1.2 云计算发展历史 ·· (2)
  1.1.3 云计算优势 ·· (3)
 1.2 云计算体系架构 ·· (4)
  1.2.1 基础设施即服务 IaaS ·· (5)
  1.2.2 平台即服务 PaaS ·· (5)
  1.2.3 软件即服务 SaaS ·· (6)
 1.3 云计算平台 OpenStack 介绍 ··· (6)
  1.3.1 OpenStack 简介 ·· (7)
  1.3.2 OpenStack 体系结构及服务组件 ·· (8)
 1.4 经典云计算解决案例 ··· (9)
  1.4.1 沃尔玛用 OpenStack 做到"天天低价" ·· (10)
  1.4.2 PayPal：8500 台服务器规模变身为最大金融 OpenStack 云 ················· (11)
  1.4.3 联想集团基于 OpenStack 构建私有云的转型实践 ··························· (12)
 课后习题 ·· (13)

第2章 虚拟化技术 ·· (14)
 2.1 虚拟化技术简介 ·· (14)
  2.1.1 虚拟化介绍 ·· (14)
  2.1.2 虚拟化分类 ·· (15)
  2.1.3 云计算时代下的虚拟化技术 ··· (16)
  2.1.4 KVM 介绍 ·· (17)
 2.2 安装和使用 ·· (17)
  2.2.1 环境准备 ··· (18)
  2.2.2 安装 KVM ·· (19)
 2.3 虚拟机管理 ·· (21)
  2.3.1 创建虚拟机 ·· (21)

	2.3.2　管理虚拟机	(24)
课后习题		(25)

第 3 章　OpenStack 环境准备 (26)

3.1	OpenStack 回顾	(26)
3.2	准备工作	(28)
	3.2.1　OpenStack 环境部署	(28)
	3.2.2　安全配置	(29)
	3.2.3　网络配置	(31)
	3.2.4　配置主机映射	(33)
	3.2.5　配置 yum 源	(34)
	3.2.6　安装 NTP 服务	(36)
	3.2.7　安装 OpenStack 包	(38)
	3.2.8　安装并配置 SQL 数据库	(38)
	3.2.9　安装并配置消息服务器	(41)
	3.2.10　安装 Memcached	(42)
课后习题		(43)

第 4 章　认证服务 Keystone (44)

4.1	Keystone 基本概念	(44)
4.2	Keystone 数据库操作	(45)
4.3	安装并配置 Keystone	(45)
4.4	配置 Apache 服务	(47)
4.5	创建 Service 和 API Endpoints	(48)
4.6	创建 domain、project、user、role	(49)
4.7	验证 Keystone 服务	(51)
课后习题		(54)

第 5 章　镜像服务 Glance (55)

5.1	Glance 基本概念	(55)
5.2	数据库配置	(55)
5.3	创建服务凭证和 API 端点	(56)
5.4	安装并配置 Glance	(57)
5.5	验证 Glance 服务	(59)
5.6	制作 CentOS 7 镜像	(60)
课后习题		(65)

第 6 章　计算服务 Nova (66)

6.1	Nova 架构及原理	(66)
6.2	安装并配置控制节点	(68)
	6.2.1　数据库配置	(68)
	6.2.2　创建服务凭证和 API 端点	(69)
	6.2.3　安装并配置 Nova 组件	(70)
6.3	安装并配置计算节点	(72)

		6.3.1　安装并配置 Nova 组件	（72）
		6.3.2　检查主机是否支持虚拟机硬件加速	（73）
	6.4	验证 Nova 服务	（74）
	课后习题		（74）
第 7 章	网络部署服务 Neutron		（75）
	7.1	Neutron 基本概念	（75）
	7.2	安装并配置控制节点	（76）
		7.2.1　数据库配置	（76）
		7.2.2　创建服务凭证和 API 端点	（76）
		7.2.3　安装并配置 Neutron 组件	（78）
	7.3	安装并配置计算节点	（82）
	7.4	验证 Neutron 服务	（83）
	课后习题		（85）
第 8 章	对象存储服务 Swift		（86）
	8.1	Swift 基本概念	（86）
	8.2	控制节点环境配置	（86）
	8.3	控制节点安装并配置 Swift	（88）
	8.4	存储节点安装并配置 Swift	（89）
		8.4.1　环境准备	（90）
		8.4.2　安装并配置	（91）
	8.5	创建并分发 Ring	（93）
		8.5.1　创建账户 Ring	（93）
		8.5.2　创建容器 Ring	（94）
		8.5.3　创建对象 Ring	（95）
		8.5.4　完成安装	（96）
	8.6	校验安装	（98）
	课后习题		（99）
第 9 章	Web 服务 Dashboard		（100）
	9.1	Dashboard 基本概念	（100）
	9.2	安装并配置 Dashboard	（100）
	9.3	验证 Dashboard 服务	（101）
	课后习题		（103）
第 10 章	块存储服务 Cinder		（104）
	10.1	Cinder 基本概念	（104）
	10.2	安装并配置控制节点	（105）
		10.2.1　数据库配置	（105）
		10.2.2　创建服务凭证和 API 端点	（105）
		10.2.3　安装并配置 Cinder 组件	（108）
	10.3	安装并配置存储节点	（110）
		10.3.1　安装工具包	（110）

　　　　10.3.2　安装并配置组件 ························· (111)
　　10.4　验证 Cinder 服务 ····························· (112)
　　课后习题 ··· (112)

第 11 章　编配服务 Heat (114)

　　11.1　Heat 基本概念 ······························· (114)
　　11.2　数据库配置 ····································· (115)
　　11.3　创建服务凭证和 API 端点 ··············· (115)
　　11.4　安装并配置 Heat ···························· (118)
　　11.5　验证 Heat ······································ (120)
　　课后习题 ··· (120)

第 12 章　运行云主机 (121)

　　12.1　创建云网络 ····································· (121)
　　　　12.1.1　Provider Network ··················· (121)
　　　　12.1.2　Self-service Network ·············· (122)
　　　　12.1.3　验证网络 ······························· (124)
　　12.2　创建云主机 ····································· (125)
　　　　12.2.1　设置密钥对 ···························· (125)
　　　　12.2.2　添加安全规则 ························· (125)
　　　　12.2.3　创建云主机 ···························· (126)
　　课后习题 ··· (128)

第 13 章　OpenStack 典型架构实现 (129)

　　13.1　OpenStack 架构及规划 ···················· (129)
　　13.2　环境准备 ·· (130)
　　13.3　Ceph 集群部署及配置 ······················ (132)
　　　　13.3.1　Ceph 的相关知识 ···················· (132)
　　　　13.3.2　Ceph 目标 ····························· (132)
　　　　13.3.3　Ceph 架构 ····························· (133)
　　　　13.3.4　Ceph 组件 ····························· (134)
　　　　13.3.5　Ceph 的地位和未来 ················· (136)
　　　　13.3.6　Ceph 的搭建 ·························· (136)
　　13.4　OpenStack 搭建 ······························ (138)
　　　　13.4.1　安装数据库 ···························· (138)
　　　　13.4.2　安装消息队列服务 ··················· (138)
　　　　13.4.3　安装 Memcached 服务 ············· (139)
　　　　13.4.4　安装认证服务 ························· (139)
　　　　13.4.5　安装镜像服务 ························· (141)
　　　　13.4.6　在控制节点安装 Cinder 服务 ····· (144)
　　　　13.4.7　安装计算服务 ························· (147)
　　　　13.4.8　在控制节点安装 Neutron 相关服务 ··· (152)

13.4.9　在计算节点安装 Neutron 相关服务 ……………………………………（157）
　　13.4.10　安装 Dashboard ……………………………………………………（161）
　13.5　OpenStack 运维案例 ………………………………………………………（162）
课后习题参考答案 ……………………………………………………………………（165）

第1章 云计算概述

1.1 云计算简介

从 20 世纪 40 年代世界上第一台电子计算机诞生至今,已经过去了半个多世纪。在这几十年里,计算模式经历了单机、终端—主机、客户端—服务器几个重要时代,发生了翻天覆地的变化。在过去的几十年里,互联网将全世界的企业与个人连接了起来,并深刻地影响着每个企业的业务运作及每个人的日常生活。用户对互联网内容的贡献空前增加,软件更多地以服务的形式通过互联网被发布和访问,而这些网络服务需要海量的存储和计算能力来满足日益增长的业务需求。

互联网使得人们对软件的认识和使用模式发生了潜移默化的改变。计算模式的变革必将带来一系列的挑战。如何获取海量的存储和计算资源?如何在互联网这个无所不包的平台上更经济地运营服务?各种新的 IT 技术对各行业将会产生怎样的影响?如何才能使互联网服务更加敏捷、更随需应变?如何让企业和个人用户更加方便、透彻地理解与运用层出不穷的服务?"云计算"正是顺应这个时代大潮而诞生的信息技术理念。目前,无论是信息产业的行业巨头还是新兴科技公司,都把云计算作为企业发展战略中的重要组成部分。云计算的号角已经吹响,势不可当。本章将解释什么是云计算,包括它的发展历史与特征优势,阐述云计算的体系结构,介绍最近趋势火热的开源项目 OpenStack,最后展示一些经典的云计算解决案例。

1.1.1 云计算概念与特征

云计算无疑是最近各大厂商纷纷追逐的对象,Salesforce.com 在 2008 年年初推出了随需应变平台 DevForce,提供一套全面的云计算架构;在洛杉矶举办的微软专业开发者大会上,微软推出了备受期待的云计算平台 Windows Azure,以提供微软各大软件的网络版本应用;Sun 实施云计算 Insight 挑战 Live Mesh;IBM 在中国无锡太湖新城科教产业园为中国的软件公司建立了第一个云计算中心……如今只要搜索"云计算",就会出现数不胜数的信息,对云计算的定义也有多种说法。对于到底什么是云计算,至少可以找到 100 种解释,很多学者和机构都对云计算赋予了不同的比喻和内涵。

维基百科认为云计算是一种能够将动态伸缩的虚拟化资源通过互联网以服务的方式提供给用户的计算模式,用户不需要知道如何管理那些支持云计算的基础设施。

Whatis.com 认为云计算是一种通过网络连接来获取软件和服务的计算模式,云计算使用户可以获得使用超级计算机的体验,用户通过笔记本电脑与手机上的瘦客户端接入云中获取需要的资源。

中国云计算专委会认为,云计算最基本的概念是:通过整合、管理、调配分布在网络各处的计算资源,并以统一的界面同时向大量用户提供服务。借助云计算,网络服务提供者可以在瞬息之间,处理数以千万计甚至亿计的信息,实现和超级计算机同样强大的效能,同时,用户可以按需计量地使用这些服务,从而实现让计算成为一种公用设施来按需而用的梦想。

国家标准与技术研究院(NIST)定义云计算是一种按使用量付费的模式,这种模式提供可用的、便捷的、按需的网络访问,进入可配置的计算资源共享池(资源包括网络、服务器、存储、应用软件、服务),这些资源能够被快速地提供,只需投入很少的管理工作,或与服务供应商进行很少的交互。

总的来说,在云计算中,IT 业务通常运行在远程的分布式系统上,而不是在本地计算机或者单个服务器上。这个分布式系统由互联网相互连接,通过开放的技术和标准把硬件和软件抽象为动态可扩展、可配置的资源,并对外以服务的形式提供给用户。该系统允许用户通过互联网访问这些服务,并获取资源。服务接口将资源在逻辑上以整合实体的形式呈现,隐蔽其中的实现细节。该系统中业务的创建、发布、执行和管理都可以在网络上进行,而用户只需要按资源的使用量或者业务规模付费。好比是从古老的单台发电机模式转向了电厂集中供电的模式。它意味着计算能力也可以作为一种商品进行流通,就像煤气、水电一样,取用方便,费用低廉。最大的不同是,它是通过互联网进行传输的。

云计算是并行计算(Parallel Computing)、分布式计算(Distributed Computing)和网格计算(Grid Computing)的发展,或者说是这些计算科学概念的商业实现。云计算是虚拟化(Virtualization)、效用计算(Utility Computing)、将基础设施作为服务 IaaS(Infrastructure as a Service)、面向服务的架构(SOA)等概念混合演进并跃升的结果。

1.1.2 云计算发展历史

云计算主要经历了 4 个阶段才发展到现在这样比较成熟的水平,这 4 个阶段依次是电厂模式、效用计算、网格计算和云计算。

(1)电厂模式阶段:电厂模式就好比利用电厂的规模效应来降低电力的价,并让用户使用起来更方便,并且不需维护和购买任何发电设备。

(2)效用计算阶段:在 1960 年左右,计算设备的价格是非常高昂的,远非普通企业、学校和机构所能承受,所以很多人产生了共享计算资源的想法。1961 年,人工智能之父麦肯锡在一次会议上提出了"效用计算"这个概念,其核心借鉴了电厂模式,具体目标是整合分散在各地的服务器、存储系统以及应用程序来共享给多个用户,让用户能够像把灯泡插入灯座一样来使用计算机资源,并且根据其所使用的量来付费。但由于当时整个 IT 产业还处于发展初期,很多强大的技术还未诞生，比如互联网等，所以虽然这个想法一直为人称道，但是总体而言"叫好不叫座"。

（3）网格计算阶段：网格计算研究如何把一个需要非常巨大的计算能力才能解决的问题分成许多小的部分，然后把这些部分分配给许多低性能的计算机来处理，最后把这些计算结果综合起来攻克大问题。可惜的是，由于网格计算在商业模、技术和安全性方面的不足，使得它并没有在工程界和商业界取得预期的成功。

（4）云计算阶段：云计算的核心与效用计算和网格计算非常类似，也是希望 IT 技术能像使用电力那样方便，并且成本低廉。但与效用计算和网格计算不同的是，2014 年在需求方面已经有了一定的规模，同时在技术方面也已经基本成熟了。

目前，云计算革命正处于高速发展阶段。全球各大 IT 巨头都倾注巨资围绕云计算展开了激烈角逐。Google 在云计算方面已经走在众多 IT 公司的前面，它对外公布的云计算科技主要有 MapReduce、GFS（Google 文件系统，Google File System）及 BigTable。从 2007 年开始，微软公司也在美国、爱尔兰、冰岛等地投资数 10 亿美元建设其用于"云计算"的"服务器农场"，每个"农场"占地都超过 7 个足球场，集成数 10 万台计算机服务器田。IBM 的蓝云计算平台是一个企业级的解决方案，它为企业客户搭建分布式、可通过互联网访问的云计算体系，整合了 IBM 自身的 Tivoli、VMware 的虚拟化软件以及 Hadoop 开源分布式文件系统，由数据中心、管理软件、监控软件、应用服务器、数据库以及一些虚拟化的组件共同组成。亚马逊的云计算名为 AmazonWeb Services（亚马逊网络服务），目前主要由 4 块核心服务组成：Simple Storage Service（简单的存储服务，S3）、Elastic Compute Cloud（弹性计算云，EC2）、Simple Queuing Service（简单排列服务）、Simple DB。其他（如雅虎、Sun 和思科等）公司围绕"云计算"也都有重大举措。

1.1.3 云计算优势

云计算的特点和优势是：快速满足业务需求；低成本，绿色节能；提高了资源利用和管理效率。云计算极大地提高了互联网应用的用户体验度，同时具备极低的成本。本节会从三个方面详细阐述云计算的优势。

1. 快速满足业务需求

（1）轻松、快速地获取服务

公有云使用者，如中小企业，可直接通过网络购买服务，省去了购买软硬件和开发的环节，企业再也不需要将精力放在应该购买什么设备，应该怎么布线，应该什么时候更新软件这些和业务完全不相干的事情上了，所有的时间、精力和资金可以完全投入业务中去，"好钢用在刀刃上"，云计算为企业的发展提供了极大的帮助。企业私有云提供的资源服务流程可为企业业务上线提供及时的资源支持。

（2）灵活、可扩展

云计算提供的资源是弹性可扩展的，可以动态部署、动态调度、动态回收，以高效的方式满足业务发展和平时运行峰值的资源需求。众所周知，企业的规模是逐渐变大的，客户的数量是逐渐增多的，随着客户的增多，访问量也急速膨胀，但是应用并不会变慢也不会堵塞，这些都归功于云服务商不断为其提供更多的存储空间和更快速的处理能力。当然，网络使用量也不是每时每刻都保持一致的，从晚上 12：00 之后到第二天上午这段时间除了"夜猫子"之外，基本上很少有人上网，而在晚上19：00～22：00的黄金时段，网络使用量又会达到峰值，"云"里的资源都可以动态分布，人多的时候，调配来的资源也会相应增多，不会浪费，也绝对不会难以满足需求。

2. 低成本、绿色节能

在海量数据处理等场景中，云计算以PC集群分布式处理方式替代小型机加盘阵的集中处理方式，可有效地降低建设成本。在激烈的商战中，赚钱当然是第一位，然而省钱也是另一种"生财之道"。Google 中国区前总裁李开复曾说，如果没有云计算技术，Google 每年购买设备的资金将高达640亿美元，而采用云计算技术后则为16亿美元。也就是说，Google只用了竞争对手1/40的成本，Google使用云存储后的存储成本是对手的1/30。

云计算通过虚拟化提高设备利用率，整合现有应用部署，降低设备数量规模。千千万万台计算机都是开着的，但真正的使用率又是多少，我们可能只是开着计算机听歌，或者仅仅只是在写文件，CPU的利用率都不到10%，甚至有时候我们只是开着计算机耗电而已。可以设想，如果每台计算机都在浪费自己90%的资源，那总量该是多么惊人！云计算和虚拟化结合在一起，就可以避免这样庞大的资源浪费。一台服务器可以虚拟成两个甚至更多的服务器，这听上去似乎难以理解，但这是事实。在客户眼中，似乎有处理文档服务器、邮件服务器、照片处理服务器，但其实这些都是由一台服务器完成的，它的30%的资源去处理文档了，30%的资源去处理照片了，这样，这台服务器的个人潜力得到了最大程度的挖掘。

云计算和虚拟化的结合提高了设备利用率，节省了设备数量，进而大大减少了用电量，在很大程度上促进了数据中心的绿色节能。

3. 提高资源管理效率

（1）集中化管理

云计算采用虚拟化技术使得跨系统的物理资源统一调配、集中运维成为可能。当你在华尔街的办公室里利用Google寻找周边实惠的聚餐场所时，你只是发出了这个请求，可是在庞大的Google的计算机群里，你并不知道到底是哪几台计算机为你服务的。管理员只需通过一个界面就可以对虚拟化环境中的各个计算机的使用情况、性能等进行监控，发布一个命令就可以迅速操作所有的机器，而不需要在每个计算机上单独进行操作。

（2）维护专业化

服务器和存储资源池的专业管理使维护人员可专注于特定领域的运维，有助于提高运维质量。IT部门也不再需要关心硬件技术细节，而集中在业务、流程设计上。

（3）系统部署和维护自动化程度提高

如果在云计算资源池中，以虚拟机方式部署应用，那么应用的上线、资源变更和物理设备切换等过程将更加简单高效。

1.2　云计算体系架构

作为一种新兴的计算模式，云计算能够将各种各样的资源以服务的方式通过网络交付给用户。云计算需要清晰的架构来实现不同类型的服务及满足用户对这些服务的各种需求。在云计算中，根据其服务集合所提供的服务类型，整个云计算服务集合被划分成4个层次：应用层、平台层、基础设施层和虚拟化层。这4个层次的每一层都对应一个子服务集，云计算服务层次如图 1-1 所示。

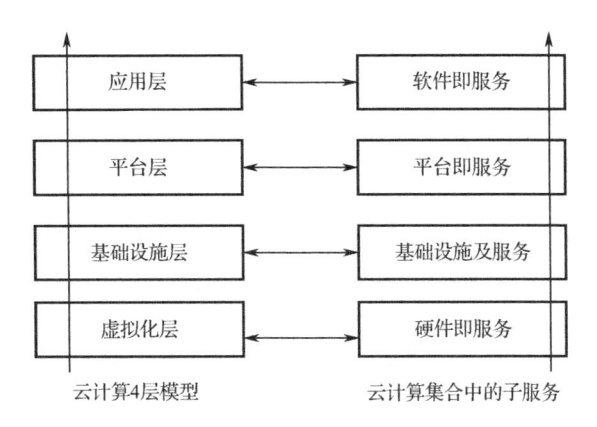

图 1-1　云计算服务层次

Sun 公司就云计算提出一个大家都比较认同的观点：云计算可描述在从硬件到应用程序的任何传统层级提供的服务。实际上，云服务提供商倾向于提供可分为如下三个类别的服务：软件即服务（Software as a Service）、平台即服务（Platform as a Service）、基础设施即服务（Infrastructure as a Service）。在云计算服务体系结构中，各层次与相关云产品对应。

1.2.1　基础设施即服务 IaaS

基础设施即服务（IaaS）交付给用户的是基本的基础设施资源。用户无须购买、维护硬件设备和相关系统软件，就可以直接在 IaaS 层上构建自己的平台和应用。基础设施向用户提供虚拟化的计算资源、存储资源和网络资源。这些资源能够根据用户的需求进行动态分配。相对于软件即服务（SaaS）和平台即服务（PaaS），基础设施即服务所提供的服务比较偏底层，但使用也更为灵活，比如，IaaS 服务可根据用户需求，提供一台装有操作系统的虚拟机，用户可用此虚拟机来运行自己的业务。全球主流 IaaS 提供商有 Amazon、Microsoft、Vmware、Rackspace 和 Red Hat。

Amazon EC2 服务是 IaaS 的典型案例。它的底层采用 Xen 虚拟化技术，以 Xen 虚拟机的形式向用户动态提供计算资源。除了 Amazon EC2 的计算资源外，Amazon 公司还提供简单存储服务（Simple Storage Service，S3）等多种 IT 基础设施服务。Amazon EC2 的内部细节对用户是透明的，因此用户可以方便地按需使用虚拟化资源。Amazon EC2 向虚拟机提供动态 IP 地址，并且具有相应的安全机制来监控虚拟机节点间的网络，限制不相关节点间的通信，从而保障了用户通信的私密性。从计费模式来看，EC2 按照用户使用资源的数量和时间计费，具有充分的灵活性。

1.2.2　平台即服务 PaaS

平台即服务（PaaS）交付给用户的是丰富的"中间件资源"，这些资源包括应用容器、数据库和消息处理等。因此，平台即服务面向的并不是普通的终端用户，而是软件开发人员，他们可以充分地利用这些开放的资源来开发定制化的应用。PaaS 公司在网上提供各种开发和分发应用的解决方案，比如虚拟服务器和操作系统。这节省了用户在硬件上的费用，也让分散的工作室之间的合作变得更加容易。一些大的 PaaS 提供者有 Google App Engine、Microsoft Azure、Force.com、Heroku、Engine Yard。最近兴起的公司有 AppFog、Mendix 和 Standing Cloud。

在 PaaS 上开发应用和传统的开发模式相比有着很大的优势。首先，由于 PaaS 提供的高级编程接口简单易用，因此软件开发人员可以在较短时间内完成开发工作，从而缩短应用上线时间；其次，由于应用的开发和运行都基于同样的平台，因此兼容性问题较少；再次，开发者无须考虑应用的可伸缩性、服务容量等问题，因为 PaaS 都已提供；最后，平台层提供的运营管理功能还能够帮助开发人员对应用进行监控和计费。

Google 公司的 Google App Engine 就是典型的 PaaS 实例。它向用户提供了 Web 应用开发平台。由于 Google App Engine 对 Web 应用无状态的计算和有状态的存储进行了有效的分离，并对 Web 应用所使用的资源进行了严格的分配，因此使得该平台上托管的应用具有很好的自动可伸缩性和高可用性。

1.2.3 软件即服务 SaaS

软件即服务（SaaS）交付给用户的是定制化的软件，即软件提供方根据用户的需求，将软件或应用通过租用的形式提供给用户使用。SaaS 大多是通过网页浏览器来接入。任何一个远程服务器上的应用都可以通过网络来运行。用户消费的服务完全是从网页（如 Netflix、MOG、Google Apps、Box.net、Dropbox 或者苹果的 iCloud）那里进入这些分类。尽管这些网页服务是用作商务和娱乐或者两者都有，但这也算是云技术的一部分。一些用于商务的 SaaS 应用包括 Citrix 的 Go To Meeting，Cisco 的 WebEx，Salesforce 的 CRM、ADP、Workday 和 SuccessFactors。

SaaS 有三个特征。第一，用户不需要在本地安装该软件的副本，也不需要维护相应的硬件资源，该软件部署并运行在提供方自有的或者第三方的环境中；第二，软件以服务的方式通过网络交付给用户，用户端只需要打开浏览器或者某种客户端工具就可以使用服务；第三，虽然 SaaS 面向多个用户，但那时每个用户都感觉是独自占有该服务。

这种软件交付模式无论是在商业上还是技术上都是一个巨大的变革。对于用户来说，他们不再需要关心软件的安装和升级，也不需要一次性购买软件许可证，而是根据租用服务的实际情况进行付费，也就是"按需付费"。对于软件开发者而言，由于与软件相关的所有资源都放在云中，开发者可以方便地进行软件的部署和升级，因此软件的生命周期不再明显。开发者甚至可以每天对软件进行多次升级，而对用户来说这些操作都是透明的，他们感觉到的只是质量越来越完善的软件服务。

另外，SaaS 更有利于知识产权的保护，因为软件的副本本身不会提供给客户，从而减少了反编译等恶意行为发生的可能。Salesforce.com 公司是 SaaS 概念的倡导者，它面向企业用户推出了在线客户关系管理软件 Salesforce CRM，已经获得了非常积极的市场反响。Google 公司推出的 Gmail 和 Google Docs 等，也是 SaaS 的典型代表。

1.3 云计算平台 OpenStack 介绍

2015 年年初，Zenoss 完成的一份名为"2014 开源云计算解析"的市场调查显示，69%的用户已经不同程度地应用云计算技术，43%的用户花费大量资源在开源技术上。在这些选择了开源云的企业中，超过 86%的企业关注 OpenStack，并且这些数值在过去几年都在不断增长。排在第二位的 CloudStack 则被远远甩在后面，只有 44%的企业关注它。至于有着悠久历史的 Eucalyptus，则在 2014 年 9 月被 HP 收购，并且在最近整合进入 Helion 云产品线，但 OpenStack 仍在该产品线上占据统治地位。毫无疑问，OpenStack 是目前最火的开源软件，超过 585 家企业，接近 4 万人通过各种方式支持着这个超过 2000 万行的开源项目的持续发展。OpenStack 一直保持着高速增长的态势。据分析，到 2018 年，OpenStack 将会拥有 33 亿美元的云市场。国内超过一半的 IaaS 提供商都在使用 OpenStack。

OpenStack 是由网络主机服务商 Rackspace 和美国宇航局联合推出的一个开源项目，目的是制定一套开源软件标准，任何公司或个人都可以搭建自己的云计算环境（IaaS），从此打破了 Amazon 等少数公司的垄断，意义非凡。下面介绍 IaaS 主流平台 OpenStack，帮助读者在进行 OpenStack 平台搭建实践前对其有一些基本的认识。

1.3.1　OpenStack 简介

OpenStack 是一整套开源软件项目的综合，它允许企业或服务提供者建立、运行自己的云计算和存储设施。Rackspace 与 NASA 是最初重要的两个贡献者，前者提供了"云文件"平台代码，该平台增强了 OpenStack 对象存储部分的功能；而后者带来了"Nebula"平台，形成了 OpenStack 其余的部分。而今，OpenStack 基金会已经有 150 多个会员，包括很多知名公司，如 Canonical、DELL、Citrix 等。

OpenStack 由几个主要的组件组合起来完成具体工作。OpenStack 支持几乎所有类型的云环境，项目目标是提供实施简单、可大规模扩展、丰富、标准统一的云计算管理平台。OpenStack 通过各种互补的服务提供了基础设施即服务（IaaS）的解决方案，每个服务提供 API 以进行集成。

OpenStack 已经走过了 7 个年头。从最初只有两个模块（服务）Nova 和 Swift 到现在已经有 20 多个模块了，每个模块作为独立的子项目开发。OpenStack 每半年发布一个版本，版本以字母顺序命名，2016 年 10 月推出了第 14 个版本 Newton，因为在编写本书之时，Newton 还没有发布 Stable 版本，所以本书选择 Newton 前一个版本 Mitaka。Mitaka 版本聚焦于可管理性、可扩展性和终端用户体验三个方面。重点在用户体验上简化了 Nova、Keynote 的使用，以及使用一致的 API 调用创建资源。Mitaka 版本中可以处理更大的负载和更为复杂的横向扩展，也是本书中所用到的版本。

OpenStack 具有下列关键特性。

（1）管理虚拟化的产品服务器和相关资源（CPU、Memory、Disk、Network），提高其利用率和资源的自动化分配（具有更高的性价比）。

（2）管理局域网（Flat、Flat DHCP、VLAN DHCP、IPv6），程序配置的 IP 和 VLAN，能为应用程序和用户组提供灵活的网络模式。

（3）带有比例限定和身份认证：这是为自动化和安全设置的，容易管理接入用户，阻止非法访问。

（4）分布式和异步体系结构：提供高弹性和高可用性系统。

（5）虚拟机镜像管理：能提供易存储、引入、共享和查询的虚拟机镜像。

（6）云主机管理：提高生命周期内可操作的应用数量，从单一用户接口到各种 API，例如一台主机虚拟的 4 台服务器，可以有 4 种 API 接口，管理 4 个应用。

（7）创建和管理云主机类型（Flavors）：为用户建立菜单使其容易确定虚拟机大小，并作出选择。

（8）iSCSI 存储容器管理（创建、删除、附加和转让容器）：数据与虚拟机分离，容错能力变强，更加灵活。

（9）在线迁移云主机。

（10）动态 IP 地址：注意管理虚拟机是要保持 IP 和 DNS 的正确。

（11）安全分组：灵活分配，控制接入云主机。

（12）按角色接入控制（RBAC）。

（13）通过浏览器的 VNC 代理：快速方便的 CLI administration。

1.3.2　OpenStack 体系结构及服务组件

OpenStack 覆盖了网络、虚拟化、操作系统、服务器等各个方面。面对如此庞大的阵容，首先介绍 OpenStack 架构（如图 1-2 所示），了解架构里哪些核心模块负责管理计算资源、网络资源和存储资源，模块之间如何协调工作，它能帮助我们站在高处看清楚事物的整体结构，避免过早地进入细节而迷失方向。本书第 13 章会构建一个实验环境，进到各个模块的内部，通过实际操作真正理解和掌握 OpenStack。

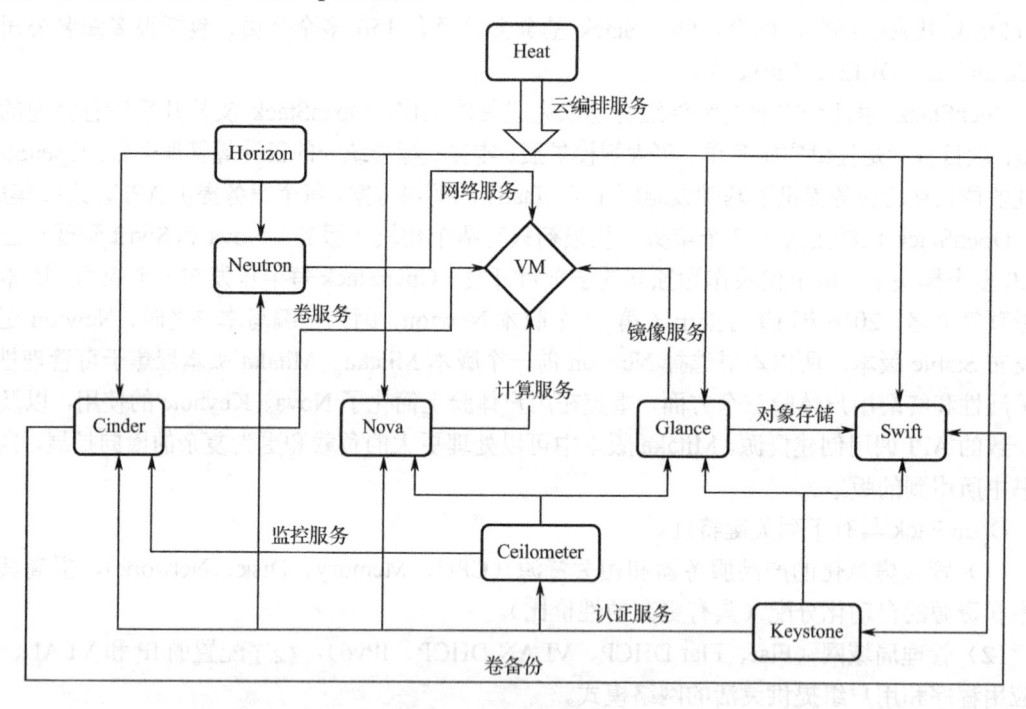

图 1-2　OpenStack 架构

VM：虚拟机（云主机），围绕 VM 的那些长方形代表 OpenStack 不同的服务。

计算（Compute）：Nova。它是最核心的。最开始的时候，Nova 可以说是一套虚拟化管理程序，还可以管理网络和存储。Nova 是一套控制器，用于为单个用户或使用群组管理云主机的整个生命周期，根据用户需求来提供虚拟服务。负责虚拟机创建、开机、关机、挂起、暂停、调整、迁移、重启、销毁等操作，配置 CPU、内存等信息规格。自 Austin 版本起集成到项目中。

对象存储（Object Storage）：Swift。一套用于在大规模可扩展系统中通过内置冗余及高容错机制实现对象存储的系统，允许进行存储或者检索文件。可为 Glance 提供镜像存储，为 Cinder 提供卷备份服务。自 Austin 版本起集成到项目中。Swift 是对象存储的组件。对于大部分用户来说，Swift 不是必需的。只有存储数量到一定级别，而且是非结构化数据才有这样的需求。Swift 是 OpenStack 所有组件里最成熟的，可以在线升级版本，各种版本可以混合在一起，也就是说，1.75 版本的 Swift 可以和 1.48 版本的 Swift 混合在一个群集里，这是很难得的。

镜像服务（Image Service）：Glance。一套虚拟机镜像查找及检索系统，支持多种虚拟机镜像格式（AKI、AMI、ARI、ISO、QCOW2、Raw、VDI、VHD、VMDK），有创建上传镜像、删除镜像、编辑镜像基本信息的功能。自 Bexar 版本起集成到项目中，目前，Glance 的最大需求就是多个数据中心的镜像管理，不过这个功能已经基本实现。还有就是租户私有的 Image 管理，这些功能目前都已经实现。

认证服务（Identity Service）：Keystone。这是提供身份认证和授权的组件。对于任何系统，身份认证和授权其实都比较复杂。尤其是对于 OpenStack 这么庞大的项目，每个组件都需要使用统一认证和授权。Keystone 为 OpenStack 其他服务提供身份验证、服务规则和服务令牌的功能，管理 Domains、Projects、Users、Groups、Roles。自 Essex 版本起集成到项目中。Keystone 还需要提供更多的功能，如基于角色的授权、Web 管理用户等。

网络服务（Network）：Neutron。提供云计算的网络虚拟化技术，为 OpenStack 其他服务提供网络连接服务。为用户提供接口，可以定义 Network、Subnet、Router，配置 DHCP、DNS、负载均衡、L3 服务，网络支持 GRE、VLAN。插件架构支持许多主流的网络厂家和技术，如 OpenvSwitch。自 Folsom 版本起集成到项目中。

块存储（Block Storage）：Cinder。这是存储管理的组件。Cinder 存储管理主要是指虚拟机的存储管理。为运行云主机提供稳定的数据块存储服务，它的插件驱动架构有利于块设备的创建和管理，如创建卷、删除卷，在云主机上挂载和卸载卷。自 Folsom 版本起集成到项目中。

UI 界面（Dashboard）：Horizon。OpenStack 中各种服务的 Web 管理门户，用于简化用户对服务的操作，使用这个 Web GUI，可以在云上完成大多数的操作，如启动云主机、分配 IP 地址、配置访问控制等。自 Essex 版本起集成到项目中。

测量（Metering）：Ceilometer。像一个漏斗一样，能把 OpenStack 内部发生的几乎所有的事件都收集起来，然后为计费、监控以及其他服务提供数据支撑。自 Havana 版本起集成到项目中。

部署编排（Orchestration）：Heat。提供了一种通过模板定义的协同部署方式，实现云基础设施软件运行环境（计算、存储和网络资源）的自动化部署。自 Havana 版本起集成到项目中。

1.4 经典云计算解决案例

在云计算技术的驱动下。计算服务正从传统的"高接触、高成本、低承诺"的服务配置向"低接触、低成本、高承诺"转变。如今，包括 IaaS、PaaS、SaaS 等模式的云计算凭借其优势获得了在全球市场的广泛认可。企业、政府、军队等各种重要部门都正在全力研发和部署云计算相关的软件和服务，云计算已进入国计民生的重要行业。本节介绍现实中应用云计算平台 OpenStack 成功进行 IT 改革的经典案例，以帮助读者对云计算有一个更全面的认识。

1.4.1 沃尔玛用 OpenStack 做到"天天低价"

沃尔玛一直通过采用先进技术推动企业发展。它是最早向供应商开放库存系统的企业之一。它还是在互联网出现之前第一家使用卫星通信连接商店网络的公司。如今，它又在云计算方面投入了大量资金。2014 年 8 月，沃尔玛将整个电子商务栈搬到在 Canonical 公司的 Ubuntu Linux 操作系统上运行的 OpenStack 上。大家想到零售巨头沃尔玛时，大多会想到物美价廉的商品，或是沃尔玛提供的方便。可能不会想到沃尔玛供应链"天天低价"背后的软件正是 OpenStack。

沃尔玛是个在砖瓦型实体大楼里做生意的零售商，每年的收入达 4800 亿美元。沃尔玛快速发展，其中"沃尔玛全球电子商务（Walmart Global eCommerce）"首当其冲。沃尔玛全球电子商务每年的增长速度超过 30%，同时，若要保持供应链的运行并能以非常低的价格提供商品，沃尔玛就需要有软件可以在 27 个国家内每星期跟踪 11000 家店里的 2.45 亿个客户。而且，沃尔玛也在迈向电子商务 3.0，该公司拥有 11 个电子商务网站，2014 年黑色星期五周末的浏览量达 15 次，这些均由电子商务网站处理。其客户希望，在家用计算机、手机、平板，甚至沃尔玛零售商店内的查询机上使用沃尔玛的电子商务平台时能获得相同的体验。

为了满足这种需求，沃尔玛需要的技术堆栈规模上必须：具有可扩展性，以期能满足爆炸性的需求；具有足够的灵活性，以构建应用程序适应不断变化的用户喜好；具有足够的大数据智慧，以预测客户想要的东西和为客户提供建议。在沃尔玛看来，使用云计算不仅能使用大量的商用机器代替价格昂贵的大型机器，还可以大大降低基础设施成本，云的分布式架构提供了更高弹性和可靠性。于是，沃尔玛决定构建一个弹性云，使用面向服务的架构运行应用程序。对于云平台的选择，沃尔玛希望云平台可以使其能够快速构建所有类型的应用程序，包括移动应用、Web 应用和 RestFul API；使产品经理能够以敏捷方式迭代；使沃尔玛能够更高效地响应客户需求。

经过长时间的考虑后，沃尔玛在 2014 年 8 月将该项技术的赌注压在 OpenStack 上。沃尔玛当时将整个电子商务栈都搬到在 Canonical 公司 Ubuntu Linux 操作系统上运行的 OpenStack 上。沃尔玛选择 OpenStack 作为其云平台，不仅是因为 OpenStack 是同类技术中最出色的，而且也因为开源软件有其与生俱来的几大优势，比如开源意味着可以修改和定制，从而便于满足沃尔玛全球电子商务的个性化需求。最重要的，使用 OpenStack 的最大优势是使沃尔玛避免了长期锁定在某一个专有供应商身上。

从开始使用 OpenStack 的 9 个月里，沃尔玛已经在超过 15 万个核心应用里建立了一个 OpenStack 计算层，这个数字还在不断上升。沃尔玛还利用 OpenStack 项目里诸如 Neutron 和 Cinder 的软件，将更多的块存储和风险项目加到软件定义网络里。沃尔玛目前还在用 Swift 建一个多 PB 级的对象存储。到 2015 年旅游旺季，沃尔玛将 OpenStack 云搬到了 2014 年的 Juno 发布版里。虽然很多人都在使用 OpenStack，但沃尔玛 OpenStack 项目令人兴奋之处在于其使用规模。他们是在真实的生产负载中使用 OpenStack 平台，而且到目前为止，Walmart.com 整个美国的流量都由该平台支撑。

有的人认为，OpenStack 还不够成熟，不足以在商业产品环境中使用，但世界上最大的零售商押上了老本，笃定地认为 OpenStack 可以挑大梁了。

1.4.2　PayPal：8500 台服务器规模变身为最大金融 OpenStack 云

全球在线支付解决方案领导者 PayPal 目前已经结束了为期三年的从传统的混合企业数据中心向 OpenStack 私有云的迁移工作。经过三年的迁移后，PayPal 表示几乎已经把所有的运营都部署在了 OpenStack 云上，包括近 100%的 PayPal 流量服务、Web/API 应用和中间层服务。2014 年，PayPal 在其基础设施中处理了金额高达 2280 亿美元的支付交易，这标志着其基础设施已经成为全球最大的、已经投入使用的金融服务 OpenStack 云。

PayPal 在 2011 年就希望对数据中心基础设施进行改造。随着 2013 年第二季度即将结束，PayPal 占 eBay 42%的收入还在持续增长，云计算的创新有助于其扩大规模，在竞争中保持领先地位。1.32 亿个活跃注册账户，支持 25 种货币支付，可用性、敏捷性和安全性，这些对 PayPal 的基础设施来说都至关重要。PayPal 的目标在不影响可用性或者损害客户对 PayPal 信任的前提下，实现大规模的敏捷性和高可用性。这就意味着 PayPal 需要将所有现在手动完成的一切纳入"即服务"中。也就是说，PayPal 将实现软件定义的 API，并且在未来 2~3 年内，这些 API 都将打包在软件定义数据中心的保护伞下。

当时，OpenStack 还只是一个半成品。借助于 VMware 虚拟化，它们成为一个自动化程度更高的基础设施。OpenStack 在研发初期就已经获得了 PayPal 的关注。PayPal 最后选择 OpenStack 的原因是因为其开放的标准和生态系统的势头。OpenStack 提供了相当强大的 API 和抽象概念。除了给 PayPal 带来这些抽象层之外，从行业领先企业角度，PayPal 也看到了 OpenStack 良好的发展前景，例如 IBM、惠普、红帽。所有重要的供应商都积极采纳 OpenStack 技术。因此，PayPal 认为 OpenStack 对他们来说相当有价值。

向 OpenStack 的迁移并不仅仅是一个基础设施的更替，也是一个企业内部文化的调整，其 IT 人员所做的调整工作已经远远超越了服务器配置范畴。PayPal 在其 OpenStack 中运行着 8500 台标准化的 x86 服务器，向 1.62 亿个客户提供信息、移动应用支持、网站交互和支付处理等服务。无状态交互（例如响应客户信息请求的 PayPal 前端界面）和状态交互（例如接收客户提交信息的后端数据库）都已在 OpenStack 上处理。目前，PayPal 也已经在 OpenStack 升级方面建立了一整套包括成立指挥中心和任命升级程序主管在内的流程与规定。为了保持 8500 台服务器的同质性，防止这些服务器使用不同的 OpenStack 版本，建立统一的表单和采取整体行动非常重要。这意味着服务器、架顶式交换机、防火墙、负载均衡器和存储器等 180000 个数据中心部件都将成为 PayPal OpenStack 云的一部分。

PayPal 的 OpenStack 云能够容纳机械故障，直至启动解决和更换所有故障设备的例行性维护工作，在维护中，这些设备可以在线更换。过去，1%的设备出现故障后，技术人员就需要进行维护，如今，这一上限值已经提升到了 3%~5%，满足了定期维护的要求。这样可以让 IT 部门以例行性和自动化方式运行 PayPal 数据中心。过去，如果数据中心服务器、交换机或存储器出现故障，那么通常的做法是尽快派人去解决这些问题。在 OpenStack 中，处理故障设备的做法是切换至状态良好的设备。

PayPal 的 OpenStack 云还具有自动感知机制，能够检测到硬件发生故障或即将发生故障的时间。自动化运行的主要目的是，在 PayPal 开发小组需要服务器时，可以迅速地为他们提供服务。在瞬息万变的移动支付领域，PayPal 通过允许大批应用频繁升级的方式紧跟需求的变化。环境的变化非常频繁，如果不迁移到 OpenStack 基础设施，每天为软件打补丁和升级的紧凑工作基本上是不可能完成的。基础设施的同质性和运行状态的可预测性，使应对软件的频繁调整

1.4.3 联想集团基于 OpenStack 构建私有云的转型实践

在以 AWS、Google、阿里等为代表的公有云发展的同时，很多大型企业出于数据安全性、系统稳定性、软硬件自主权、对自主可控以及 TCO 低的考虑，更加倾向于建设企业私有云来承载内部业务信息系统的运行。在中国市场，说得上的 OpenStack 案例，往往来自互联网公司，尽管有企业就绪的呼声，传统行业对 OpenStack 仍然观望居多。IT 大厂联想集团基于 OpenStack 构建私有云的转型实践，从技术架构和业务应用层面，验证了 OpenStack 的企业级高可用性，这是难能可贵的一大步。

联想以往的内部 IT 主要以面向大型客户以及渠道为主，系统架构以包括 IBM Power 小机、AIX、PowerVM、DB2 在内的及近年普遍使用的 VMware 虚拟化的传统 IT 架构构建而成。在向互联网企业转型的过程中，首先用户数与交易量就远远无法支撑全新的 B2C 的业务。其次，在成本方面，联想以往的 IT 架构是大规模采用商用化的解决方案，虽可靠但不便扩展且成本昂贵。此外，对于 IT 团队的效率与安全合规性，传统的 IT 架构仍然无法支撑起联想面向电商与移动新业务转型。在走向移动化、社交网络的过程中，无论传统的 PC 与手机都经历着激烈的竞争及快速的技术转变。在面临市场的飞速演变与竞争中，联想集团提出了从产品导向到用户导向转型的新战略。而只有可快速迭代、弹性扩展的企业云平台才能够支撑联想这种业务创新的需求。

在选型过程中，联想在对主流的 x86 虚拟化技术、私有云平台、公有云进行了全面分析与对比后，从稳定性、可用性、开放性，以及生态系统的全面与活跃度等因素考虑，最终认为 OpenStack 云平台技术可以满足联想的企业需求，确定采用 OpenStack 作为其业务持续创新的基础云平台。经过半年多的实践，联想已经建设成为规模超过 3000 Core 的 OpenStack 生产级环境，数据以最高 10TB/天的速度快速增长，并在 2015 年内将近 20% IT 负载迁移到了云环境。

在逻辑架构上，联想企业云平台完全通过软件定义环境的方式来管理基础架构，底层采用 x86 服务器以及 10Gb 网络，引入互联网式的监控运维解决方案，并用 OpenStack 平台来管理所有资源，如图 1-3 所示。

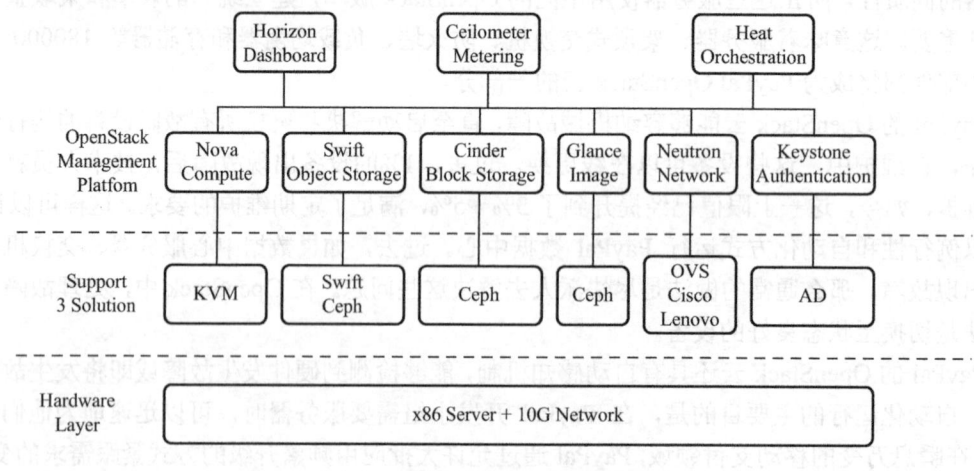

图 1-3 联想企业云平台逻辑架构

为了将 OpenStack 提升至企业级服务水平，联想在计算、网络、存储等方面战胜了很多挑战。

在计算方面，联想采用高密度的虚拟机部署方式，底层基于 KVM 虚拟化技术，通过多种优化手段，发挥物理机最大性能，在计算存储融合架构下对 CPU、内存等硬件资源做隔离。最终实现在每台双路 CPU 计算节点上保证 50 台以上虚拟机仍能平稳高效运行。另外，在云环境中一般提倡应用程序自身高可用性来应对硬件故障，但仍然有一些应用属于传统应用，对于单个主机的可用性还有要求。对于不能实现高可用性的传统应用，联想通过 Compute HA 技术实现了计算节点的高可用性，通过多种检测手段判定计算节点是否发生故障，将故障物理机上的虚拟机迁到其他可用的物理机上，整个过程无人值守，最大限度地减少因为物理机故障导致的业务中断。

在网络方面，使用不同网卡、不同交换机或不同 VLAN 将各种网络隔离，以避免网络相互干扰，达到提高整体带宽和更好地监控网络的目的。通过多个 Public 网络实现网络灵活性，便于管理安全策略。使用 VLAN 网络模式，与传统数据中心网络更好地整合，通过优化 VLAN 数据包处理，达到很好的网络数据包处理能力，让虚拟机网络带宽接近物理网络带宽。通过双网卡绑定到不同的交换机达到物理网络的高可用性。通过多个网络节点，实现公网的负载均衡及 HA，实现高性能和高可用性，网络节点使用 Router 级别的 Active/Standby 方式实现 HA，使用独立的网络路由监控服务确保网络 HA 的稳定性。

在存储方面，联想 OpenStack 云平台采用 Ceph 作为统一存储后端，其中 Glance 镜像、Nova 虚拟机系统盘、Cinder 云硬盘的数据存储由 Ceph RBD 提供，利用 Ceph 的 Copy on Write 特性，通过修改 OpenStack 代码，可做到秒级虚拟机部署。Ceph 作为统一存储后端，其性能无疑是企业核心应用是否虚拟化、云化的关键指标之一。在计算存储共同运行的超融合部署架构中，存储性能调优既要最大化存储性能，又要保证计算和存储资源的隔离，保证系统的稳定性。针对整个 IO 栈，联想从下往上，又对各层进行了优化。

此外，联想还将自身的业务需求融入了 OpenStack 企业云平台中，作为一个拥有数万名员工的大企业，需要通过 AD 活动目录来进行认证，员工就不用再单独建用户、记口令等；通过协作方的定制开发，联想已将 AD 功能融入 OpenStack 企业云平台之中。

在构建好整个 OpenStack 企业云平台之后，联想面向"互联网"转型的关键才得以展开，电子商务、大数据分析、IM、手机在线业务支撑等互联网业务，从测试到生产真正地由联想企业云平台支撑起来。从创新应用的测试团队反馈来看，联想企业云平台目前运行良好。

课后习题

1．云计算主要经历了 4 个阶段才发展到现在这样比较成熟的水平，这 4 个阶段依次是_____、_____、_____和_____。

2．云计算的优势是：_____；_____。

3．在云计算中，根据其服务集合所提供的服务类型，整个云计算服务集合被划分成 4 个层次：_____、_____、_____和_____。这 4 个层次每一层都对应着一个子服务集。

4．IaaS、PaaS 和 SaaS 的中英文全称分别是什么？

5．从 OpenStack 架构图中可以了解到，OpenStack 的服务组件有以下 9 个，分别是_____、_____、_____、_____、_____、_____、_____、_____、_____。

第2章 虚拟化技术

2.1 虚拟化技术简介

2.1.1 虚拟化介绍

随着 IT 规模日益庞大，高能耗、数据中心空间紧张、IT 系统总体成本过高等各方面问题接踵而至。而现有服务器、存储系统等设备没有被充分地利用起来，资源极度浪费。IT 基础架构对业务需求反应不够灵活，不能有效地调配系统资源适应业务需求。企业需要建立一种可以降低成本、具有智能化和安全特性、并能够与当前的业务环境相适应的灵活、动态的基础设施和应用环境，以更快速地响应业务环境的变化，并且降低数据中心的运营成本。在这种情况下，虚拟化技术应运而生。

虚拟化（Virtualization）是一种资源管理技术，它将计算机的各种实体资源，如服务器、网络、内存及存储等，予以抽象、转换后呈现出来，打破实体结构间的不可切割的障碍，使得用户可以比原本的组态更好的方式来应用这些资源，以提高系统的弹性和灵活性，降低成本、改进服务、减少管理风险等。

对于虚拟化技术，大多数人接触的最早且最多的应该是虚拟机（Virtual Machine）。它是通过软件模拟的具有完整硬件系统功能的、运行在一个完全隔离环境中的完整计算机系统。对完整计算机系统的简单解释就是一台含有 CPU、内存、硬盘、显卡、网卡、光驱等设备的计算机，只是对于虚拟机来说，这些设备都是通过软件模拟出来的。

计算机的虚拟化使单个计算机看起来像多个计算机或完全不同的计算机，从而提高资源利用率并降低 IT 成本。而后，随着 IT 架构的复杂化和企业应用计算需求的急剧加大，虚拟化技术发展到了使计算机看起来像一台计算机以实现统一的管理、调配和监控。比如，服务器聚合或网络计算。现在，整个 IT 环境已逐步向云计算时代跨越，虚拟化技术也从最初的侧重于整合数据中心内的资源，发展到可以跨越 IT 架构实现包括资源、网络、应用和桌面在内的全系统虚拟化，进而提高灵活性。

虚拟机的出现大大提高了物理机的资源利用率，并且在同一台物理机上给不同的应用一个隔离的运行环境，更关键的是虚拟机的资源管理比物理机便利很多。需要说明的是，虚拟机相比物理机而言是有资源、性能损耗的，但是考虑到虚拟机的众多优势和大多数场景下物理机资源利用率本身就不高，以及虚拟机技术已经成熟等因素，在现今的 IT 基础架构中，虚拟机已作为一个不可或缺的部分得到广泛应用。

2.1.2 虚拟化分类

按照不同的方式，虚拟化有多种分类。

（1）按照操作系统耦合程度分类，可分为全虚拟化和半虚拟化。

① 全虚拟化（Full Virtualization）又叫硬件辅助虚拟化技术，最初所使用的虚拟化技术就是全虚拟化技术，它在虚拟机（VM）和硬件之间加了一个软件层——Hypervisor，或者称为虚拟机管理程序（VMM）。Hypervisor 可以划分为两种：一种是直接运行在物理硬件之上的，如基于内核的虚拟机（KVM——它本身是一个基于操作系统的 Hypervisor）；另一种运行在另一个操作系统（运行在物理硬件之上）中，包括 QEMU 和 WINE。因为运行在虚拟机上的操作系统通过 Hypervisor 来最终分享硬件，所以虚拟机发出的指令需经过 Hypervisor 捕获并处理。为此，每个客户操作系统（Guest OS）所发出的指令都要被翻译成 CPU 能识别的指令格式，这里的客户操作系统即运行的虚拟机，所以 Hypervisor 的工作负荷会很大，因此会占用一定的资源，从而在性能方面不如裸机，但运行速度要快于硬件模拟。全虚拟化最大的优点是，运行在虚拟机上的操作系统没有经过任何修改，唯一的限制是操作系统必须能够支持底层的硬件，因为目前的操作系统一般都能支持底层硬件，所以这个限制就变得微不足道了。

② 半虚拟化（Para Virtualization）是后来才出现的技术，也称为准虚拟化技术，现在比较热门。它是在全虚拟化的基础上，对客户操作系统进行了修改，增加了一个专门的 API。这个API 可以将客户操作系统发出的指令进行最优化，即不需要 Hypervisor 耗费一定的资源进行翻译操作，因此 Hypervisor 的工作负担变得非常小，从而整体的性能也有很大的提高。缺点是，要修改包含该 API 的操作系统。但是，对于某些不含该 API 的操作系统（主要是 Windows）来说，就不能用这种方法，Xen 就是一个典型的半虚拟化的技术。

（2）从不同的角度解决不同的问题来对虚拟化技术进行分类，可分为服务器虚拟化、桌面虚拟化、应用虚拟化等。

① 服务器虚拟化能够通过区分资源的优先次序，并随时随地能将服务器资源分配给最需要它们的工作负载来简化管理和提高效率，从而减少为单个工作负载峰值而储备的资源。通过服务器虚拟化技术，用户可以动态启用虚拟服务器（又叫虚拟机），每个服务器实际上可以让操作系统（以及在上面运行的任何应用程序）误以为虚拟机就是实际硬件。运行多个虚拟机还可以充分发挥物理服务器的计算潜能，迅速应对数据中心不断变化的需求。对数量少的情况推荐使用 ESXi、XenServer。对数量大的情况推荐使用 KVM、RHEV（并不开源）、oVirt、OpenStack、VMware vSphere。

② 桌面虚拟化依赖于服务器虚拟化，在数据中心的服务器上进行服务器虚拟化，生成大量的独立的桌面操作系统（虚拟机或者虚拟桌面），同时根据专有的虚拟桌面协议发送给终端设备。用户终端通过以太网登录到虚拟主机上，只需要记住用户名和密码及网关信息，即可随

时随地通过网络访问自己的桌面系统，从而实现单机多用户。多用于 IP 外包、呼叫中心、银行办公、移动桌面。

③ 应用虚拟化技术原理是基于应用/服务器计算 A/S 架构，采用类似虚拟终端的技术，把应用程序的人机交互逻辑（应用程序界面、键盘及鼠标的操作、音频输入/输出、读卡器、打印输出等）与计算逻辑隔离开来。在用户访问一个服务器虚拟化后的应用时，用户计算机只需要把人机交互逻辑传送到服务器端，服务器端为用户开设独立的会话空间，应用程序的计算逻辑在这个会话空间中运行，把变化后的人机交互逻辑传送给客户端，并且在客户端相应设备上展示出来，从而使用户获得如同运行本地应用程序一样的访问感受。

2.1.3　云计算时代下的虚拟化技术

现在，当整个 IT 界正处于逐渐步入云计算时代的过程中，单个虚拟化技术虽然都为企业在 IT 方面带来了收益，但是人们更看重的是基于所面对的各自不同的独特环境发展出一个适合自己的、全面的虚拟化战略。我们需要考虑的是，将所有可用的虚拟化技术作为一个整体来考虑和组合，以使从中产生的效益最大化。也就是说在云计算环境下，所有虚拟化解决方案都是集服务器、存储、网络设备、软件及服务于一体的系统整合方案，并根据不同的应用环境灵活地将若干层面组合以实现不同模式虚拟化方案。

在这种云环境下的整体虚拟化战略中，我们可以利用虚拟化技术提供的多种机制，在不需重要的硬件和物理资源扩展的前提下，通过不同的方案快速模拟不同的环境和试验，达到预先构建操作 IT 系统、应用程序，提高安全性以及实现管理环境的目的，便于以后以更为简化和有效的方式将它们投入到生产环境中，进而提供更大的灵活性，并迅速确定潜在的冲突。同时，我们可以利用服务器虚拟化技术将大量分散的、没有得到充分利用的物理服务器工作负荷整合到独立的、聚合的、数量较少的物理服务器上，甚至使一台单一的大型网络虚拟机取代数以百计甚至千计的较小服务器并使其在长时间内在高利用率下运行，从而更好地管理 IT 成本、最大化能源效率及提高资源利用率。我们还可利用存储虚拟化技术来支持网络环境下多种多样的磁盘存储系统，通过将存储容量整合到一个存储资源池中，帮助 IT 系统简化存储基础架构，对信息进行生命周期管理并维护业务持续性。当然，我们还可利用应用及桌面虚拟化技术提供应用基础设施虚拟化功能，降低创建、管理和运行企业应用程序及 SOA 环境所需的运营和能源成本，并达成提高灵活性和敏捷性，确保业务流程完整性，以改进服务，提高应用程序性能并更好地管理应用程序运行状况等目的。除此之外，虚拟化的系统管理及监控服务还能帮助我们通过一个共同的接入点发现、监控和管理包括系统和软件在内的所有的虚拟和物理资源，并提供完全的跨企业服务管理，减少支持多种类型服务器所需管理工具的数量。

虚拟化是云计算的基石，虚拟化负责将物理资源池化，而云计算是让用户对池化的物理资源进行便捷的管理，所以云计算提供的从本质上讲正是虚拟化服务。从虚拟化到云计算的过程中，我们实现了跨系统的资源动态调度，将大量的计算资源组成 IT 资源池，用于动态创建高度虚拟化的资源供用户使用，从而最终实现应用、数据和 IT 资源以服务的方式通过网络提供给用户，并以前所未见的高速和富有弹性的方式来完成任务。换句话说，我们正经历一场发生在 IT 内外的迈向云计算时代的巨大变革，而推动这场变革的正是由不断发展的虚拟化技术所带来的从组件走向层级然后走向资源池的过程。云计算是虚拟化的最高境界，虚拟化是云计算的底层结构。

2.1.4　KVM 介绍

KVM（Kernel-based Virtual Machine）是一种基于 Linux x86 硬件平台的开源全虚拟化解决方案。它依托于 CPU 虚拟化指令集，性能、安全性、兼容性、稳定性表现很好，每个虚拟化操作系统表现为单个系统进程，可与 Linux 安全模块——Selinux 安全模块很好地结合。

KVM 作为 Hypervisor，主要涵盖两个重要组成部分：一个是 Linux 内核的 KVM 模块，另外一个是提供硬件仿真的 QEMU（Quick Emulator）。另外，为了使 KVM 整个虚拟化环境能够易于管理，还需要 Libvirtd 服务和基于 Libvirt 开发出来的管理工具。KVM 架构包括 KVM 模块、QEMU、Libvirt、Libvirtd、Virsh、Virt-Manager 等。

KVM 模块的主要功能是提供物理 CPU 到虚拟 CPU 的一个映射，提供虚拟机的硬件加速来提升虚拟机的性能。KVM 模块本身无法作为一个 Hypervisor 模拟出一个完整的虚拟机，并且我们也无法直接对 Linux 内核进行操作，所以需要借助其他的软件来进行，QEMU 就扮演着一个这样的角色。

QEMU 本身就是一个宿主型的 Hypervisor，即使没有 KVM，它也可以通过模拟来创建和管理虚拟机。QEMU 又借助了 KVM 的模块来提升虚拟化的整体性能。最终两者的结合提供了众所周知的 KVM 虚拟化解决方案。最终在 CentOS/RHEL 7 里用来实现 KVM 虚拟化的软件也被赋予了一个有趣的名字——qemu-kvm。

Libvirt 是 Linux 系统下一套开源的 API，主要给 KVM 的客户端管理工具提供一套方便和可靠的编程接口，它本身使用 C 语言来编写但同时也支持多种编程语言，如 Python、Perl、Ruby 和 Java 等。Libvirt 也支持多种虚拟化平台，如 KVM、Xen、ESX 和 QEMU 等。

Libvirtd 是运行在 KVM 主机上的一个服务端守护进程，为 KVM 以及它的虚拟机提供本地和远程的管理功能，基于 Libvirt 开发出来的管理工具可通过 Libvirtd 服务来管理整个 KVM 环境。

Libvirt 是一套标准的库文件，给多种虚拟化平台提供一个统一的编程接口，相当于管理工具需要基于 Libvirt 的标准接口来进行开发，在开发完成后的工具可支持多种虚拟化平台。而 Libvirtd 是一个在 Host 主机上运行的守护进程，在管理工具和 KVM 之间起到一个桥梁的作用，管理工具可通过 Libvirtd 服务来管理整个虚拟化环境。

Virsh 是基于 Libvirt 开发的一个命令行的 KVM 管理工具，可使用直接模式（Direct Mode）或交互模式（Interactive Mode）来实现虚拟机的管理，如创建、删除、启动、关闭等。

Virt-Manager 同样也是一个 KVM 管理工具，不过它是基于图形界面的。

基于以上各个组件，我们才能够部署出一套完整的 KVM 虚拟化环境。随着虚拟化技术多年的发展到现在的成熟，很多概念性东西的边界变得并不是那么清晰。好比我们之前所介绍的全虚拟化和半虚拟化，现在，XEN 也支持全虚拟化，KVM 也支持半虚拟化。

2.2　安装和使用

之前介绍了虚拟化技术，下面介绍 KVM 的安装使用。

2.2.1 环境准备

1．硬件

CPU：Intel Core i7。

内存：8GB。

硬盘：1TB。

2．基本配置

操作系统：CentOS 7.2。

IP 地址：192.168.100.30。

软件包：CentOS 自带软件包。

关闭 Selinux。

关闭防火墙。

3．虚拟化支持

KVM 虚拟化需要 CPU 的硬件虚拟化加速的支持，在本环境中为 Intel 的 CPU，在 BIOS 中开启 Intel VT。

物理机：在 BIOS 中设置，不同品牌的计算机的设置略有不同，如图 2-1 所示。

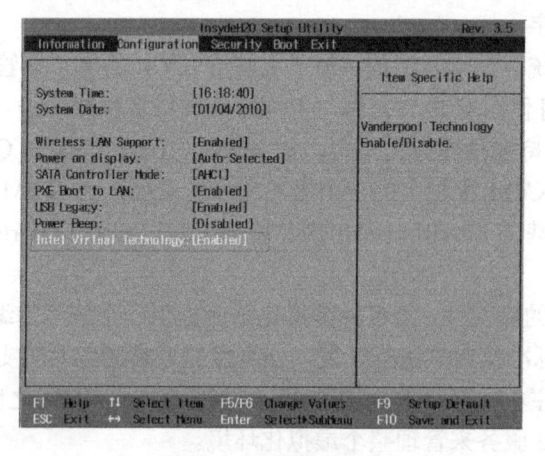

图 2-1　设置物理机

4．VMware 虚拟机

设置虚拟机，如图 2-2 所示。

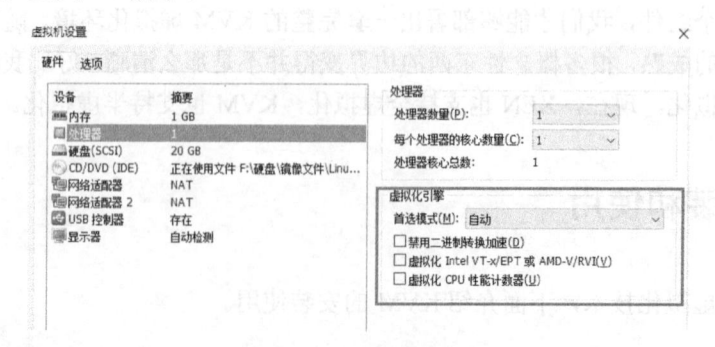

图 2-2　设置虚拟机

2.2.2 安装 KVM

软件包安装：

#yum install qemu-kvm libvirt virt-install virt-manager virt-top libguestfs-tools -y

```
Installed:
  libvirt.x86_64 0:1.2.17-13.el7
  qemu-img-ev.x86_64 10:2.3.0-31.el7_2.10.1
  qemu-kvm-common-ev.x86_64 10:2.3.0-31.el7_2.10.1
  qemu-kvm-ev.x86_64 10:2.3.0-31.el7_2.10.1
  virt-install.noarch 0:1.2.1-8.el7
  virt-manager.noarch 0:1.2.1-8.el7

Dependency Installed:
  libvirt-daemon-config-nwfilter.x86_64 0:1.2.17-13.el7
  libvirt-daemon-driver-lxc.x86_64 0:1.2.17-13.el7
  libvirt-python.x86_64 0:1.2.17-2.el7
  python-ipaddr.noarch 0:2.1.9-5.el7
  virt-manager-common.noarch 0:1.2.1-8.el7
  vte3.x86_64 0:0.36.4-1.el7

Replaced:
  qemu-img.x86_64 10:1.5.3-105.el7            qemu-kvm.x86_64 10:1.5.3-105.el7
  qemu-kvm-common.x86_64 10:1.5.3-105.el7

Complete!
[root@localhost ~]#
```

启动并设置开机启动 libvirt 服务：

 # systemctl enable libvirtd.service
 # systemctl start libvirtd.service

KVM 网络连接有以下两种方式。

（1）用户网络（User Networking）：让虚拟机访问主机、互联网或本地网络上的资源的简单方法，但是不能从网络或其他的客户机访问客户机，性能上也需要大的调整。

（2）虚拟网桥（Virtual Bridge）：网桥（bridge）模式可以让客户机和宿主机共享一个物理网络设备连接网络，客户机有自己的独立 IP 地址，可以直接连接与宿主机一样的网络，客户机可以访问外部网络，外部网络也可以直接访问客户机（就像访问普通物理主机一样）。即使宿主机只有一个网卡设备，使用 bridge 的方式也可以让多个客户机与宿主机共享网络设备，其使用非常方便，其应用也非常广泛。

1. 停止 NetworkManager 服务

 # systemctl stop NetworkManager

该服务开启的情况下修改网卡的配置文件可能会造成信息的匹配错误而导致网卡激活失败。

2. 修改网卡配置文件

备份：

cp /etc/sysconfig/network-scripts/ifcfg-enp1s0 /etc/sysconfig/network-scripts/ifcfg-enp1s0.bak
 # vi /etc/sysconfig/network-scripts/ifcfg-enp1s0

（需按照实际网卡修改。）

做如下修改：

```
TYPE=Ethernet
BRIDGE=br0
BOOTPROTO=static
NAME=enp1s0
UUID=d46f1111-3f34-481b-b3ff-e5b7e01d009a
DEVICE=enp1s0
ONBOOT=yes
```

3. 新增 br0 网桥文件并如下配置

 # vi /etc/sysconfig/network-scripts/ifcfg-br0

```
TYPE=bridge
BOOTPROTO=static
NAME=enp1s0
UUID=d46f1111-3f34-481b-b3ff-e5b7e01d009a
DEVICE=br0
ONBOOT=yes
NM_CONTROLLED=no
IPADDR=192.168.100.30
NETMASK=255.255.255.0
GATEWAY=192.168.100.1
```

NM_CONTROLLED 这个属性值，根据 RedHat 公司的文档是必须设置为 "no" 的（这个值为 "yes" 表示可以由服务 NetworkManager 来管理。NetworkManager 服务不支持桥接，所以要设置为 "no"）。但实际上发现设置为 "yes" 没有问题，通信正常。

4. 禁用网络过滤器并重新加载 kernel 参数

向文件**/etc/sysctl.conf** 添加以下代码：

vi /etc/sysctl.conf

```
# sysctl -p
net.ipv4.ip_forward = 0
net.bridge.bridge-nf-call-ip6tables = 0
net.bridge.bridge-nf-call-iptables = 0
net.bridge.bridge-nf-call-arptables = 0
```

5. 重启网络服务

 # systemctl restart network.service

重启后连接 br0 网卡 IP：192.168.100.30。

 # systemctl restart NetworkManager.service

6. 验证内核模块

 # lsmod |grep kvm

```
[root@localhost ~]# lsmod |grep kvm
kvm_intel              162153  0
kvm                    525259  1 kvm_intel
[root@localhost ~]#
```

以上输出说明内核模块加载成功，其中：
KVM 作为核心模块，协同 QEMU 实现整个虚拟化环境的正常运行。
kvm_intel 作为平台（Intel）独立模块，激活 KVM 环境的 CPU 硬件虚拟化支持。

7. 连接 Hypervisor

 # virsh connect --name qemu:///system
 # virsh list

```
[root@localhost ~]# virsh list
 Id    Name                           State
----------------------------------------------------

[root@localhost ~]#
```

这里因为没有创建虚拟机，所以显示为空，在下一节讲解虚拟机的创建及管理。

2.3 虚拟机管理

2.3.1 创建虚拟机

1. 首先通过命令 **virt-manager** 启动图形界面

```
[root@localhost ~]# virt-manager
```

2. 单击图 2-3 左端的图标创建一个虚拟机

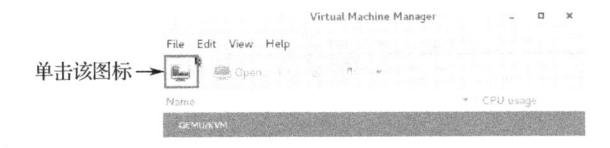

图 2-3　创建虚拟机

3. 选择第一个单选钮，全新安装一个虚拟机（如图 2-4 所示）

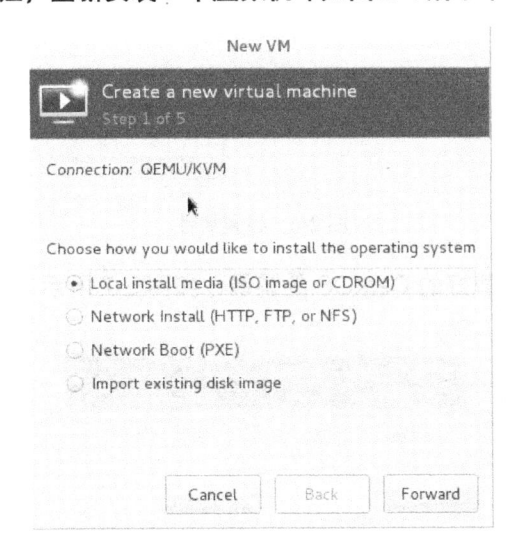

图 2-4　安装一个虚拟机

4. 单击"Browse"按钮（如图 2-5 所示）

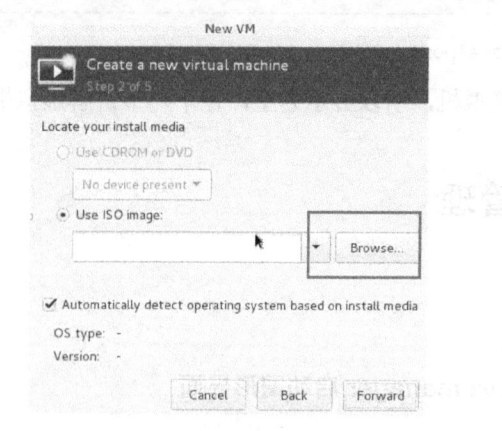

图 2-5　添加镜像

5. 单击"Browse Local"按钮（如图 2-6 所示）

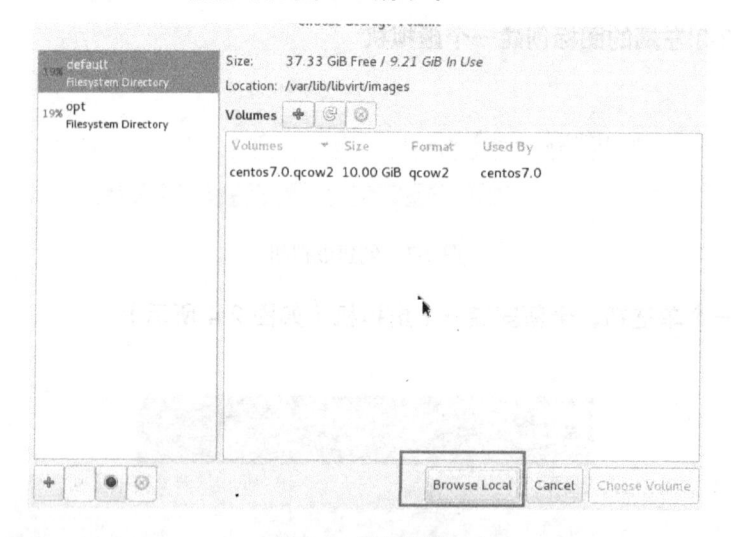

图 2-6　设置属性

6. 选择路径下已经下载好的 CentOS 的镜像（如图 2-7 所示）

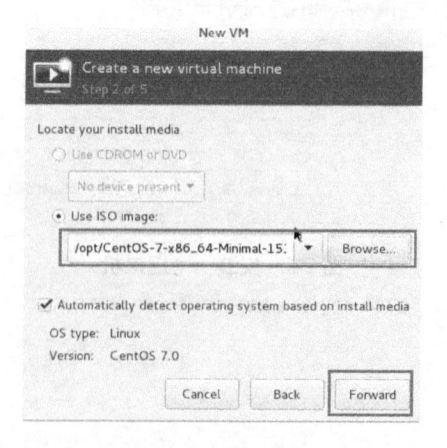

图 2-7　选择镜像

7. 配置虚拟机的内存以及 CPU（如图 2-8 所示）

图 2-8　配置参数

8. 配置磁盘大小为 9GB（如图 2-9 所示）

图 2-9　设置磁盘

9. 设置主机名为 centos7.0-2（如图 2-10 所示）

图 2-10　设置主机名

10. 单击图 2-11 左上角的 "Begin Installation"，完成安装

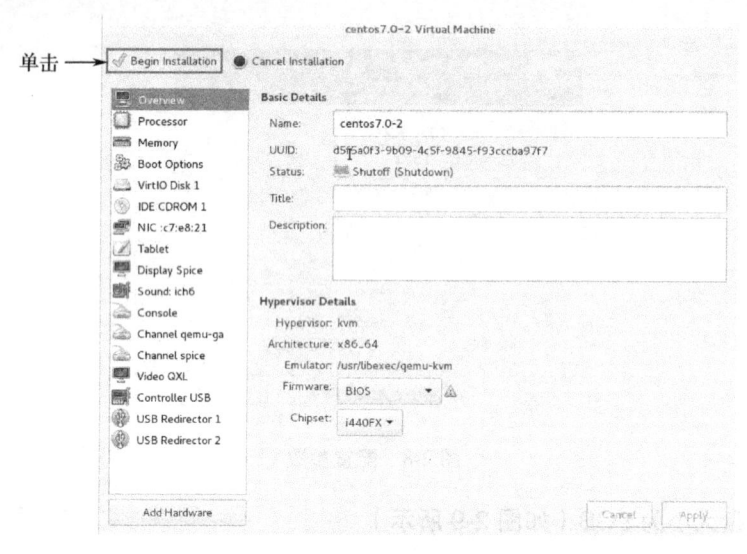

图 2-11　完成安装

2.3.2　管理虚拟机

1. 查看正在运行的虚拟机

virsh list

```
[root@localhost ~]# virsh list
 Id    Name                           State
----------------------------------------------------
 2     KVM1                           running
 4     centos7.0                      running
```

2. 查看 KVM1 这个虚拟机的详细信息

virsh dominfo KVM1

```
[root@localhost ~]# virsh dominfo KVM1
Id:             2
Name:           KVM1
UUID:           c4629455-124a-44a1-95eb-79193b9e8644
OS Type:        hvm
State:          running
CPU(s):         1
CPU time:       87.0s
Max memory:     1048576 KiB
Used memory:    1048576 KiB
Persistent:     yes
Autostart:      disable
Managed save:   no
Security model: selinux
Security DOI:   0
Security label: system_u:system_r:svirt_t:s0:c301,c563 (permissive)
```

3. 查看所有虚拟机的运行状态

virt-top

```
[root@localhost ~]# virt-top
virt-top 17:26:38 - x86_64 8/8CPU 800MHz 7717MB
2 domains, 2 active, 2 running, 0 sleeping, 0 paused, 0 inactive D:0 O:0 X:0
CPU: 0.0%  Mem: 3072 MB (3072 MB by guests)
   ID S RDRQ WRRQ RXBY TXBY %CPU %MEM    TIME   NAME
    2 R                       0.0  0.0  1:27.50 KVM1
    4 R                       0.0  0.0  1:33.21 centos7.0
```

4. 启动和关闭虚拟机

关闭虚拟机：KVM1。

 # virsh shutdown KVM1

```
[root@localhost ~]# virsh shutdown KVM1
Domain KVM1 is being shutdown
```

开启虚拟机：KVM1。

 # virsh start KVM1

```
[root@localhost ~]# virsh start KVM1
Domain KVM1 started
```

激活虚拟机自动启动：KVM1。

 # virsh autostart KVM1

```
[root@localhost ~]# virsh autostart KVM1
Domain KVM1 marked as autostarted
```

取消激活虚拟机自动启动：KVM1。

```
[root@localhost ~]# virsh autostart --disable KVM1
Domain KVM1 unmarked as autostarted
```

 # virsh autostart --disable KVM1

通过上面的 KVM 的实验，如果对虚拟机生命周期管理通过命令行来操作，在实际的生产环境中是不现实的，这就催生出了云计算平台，即一种可以快速、简单、便捷管理虚拟机生命周期的平台。当然，这只是云计算平台的一小部分功能。下面以开源云平台 OpenStack 进行云平台的学习。

课后习题

1. 按照不同的方式，虚拟化有多种分类。（1）按照操作系统耦合程度分类，可分为____和_____。（2）从不同的角度解决不同的问题来对虚拟化技术进行分类，可分为_____、_____、_____等。
2. KVM（Kernel-based Virtual Machine）是一种基于 Linux x86 硬件平台的_____。它依托于_____，性能、安全性、兼容性、稳定性表现很好，每个虚拟化操作系统表现为单个系统进程，可与 Linux 安全模块——Selinux 安全模块很好地结合。
3. KVM 网络连接有两种方式，分别是_____和_____。
4. 查看名为 KVM123 的虚拟机的详细信息的命令是_____。
5. 验证内核模块的命令是_____。

第3章 OpenStack 环境准备

3.1 OpenStack 回顾

终于正式进入 OpenStack 操作部分了。从现在开始，将带着读者一步一步地揭开 OpenStack 的神秘面纱。

OpenStack 作为 IaaS 层的一种管理平台，我们可以称之为云操作系统。对于云操作系统，如果感觉不好理解，可以设想我们平时用的笔记本、PC 等都装有操作系统，比如 Windows、Linux，正是安装了操作系统之后，我们在使用笔记本或者 PC 时不必关心系统如何调用 CPU、内存等资源，OpenStack 为虚拟机提供并管理三大类资源：计算、网络和存储（这三个就是核心），OpenStack 的核心模块如图 3-1 所示。所以我们的学习重点是：搞清楚 OpenStack 是如何对计算、网络和存储资源进行管理的。

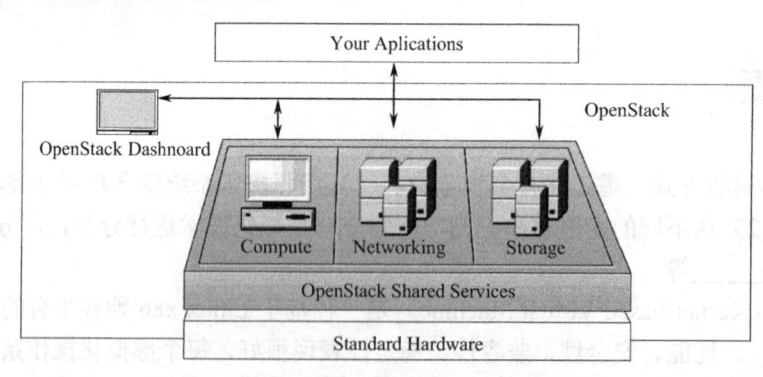

图 3-1 OpenStack 的核心模块

先回顾 OpenStack 架构（如图 3-2 所示）。它能帮助我们站在高处看清楚事物的整体结构，避免过早地进入细节而迷失方向。

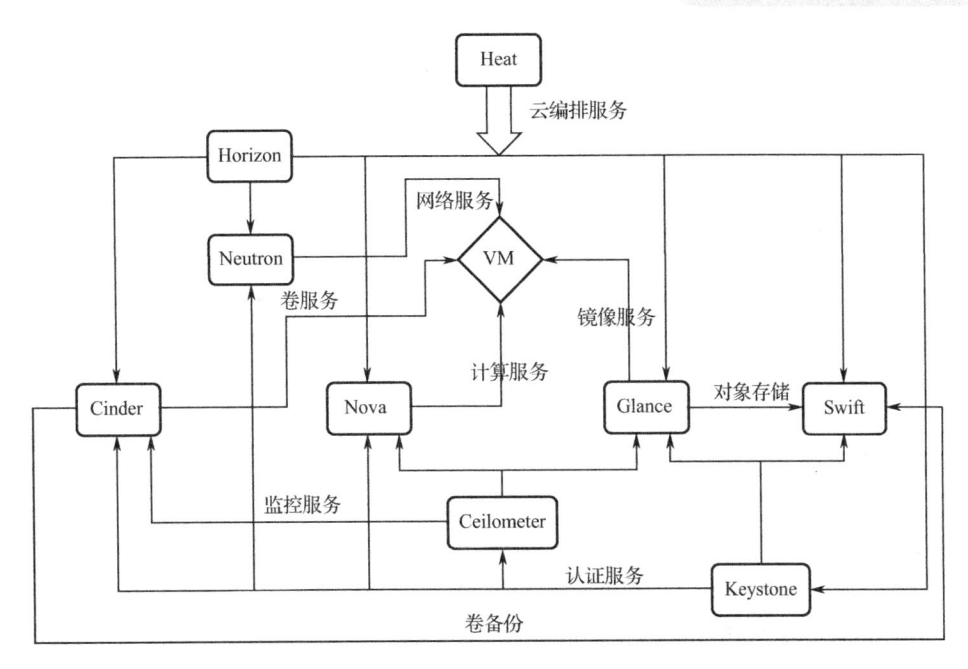

图 3-2　OpenStack 架构

VM：虚拟机（云主机），围绕 VM 的那些长方形代表 OpenStack 不同的服务。

Nova：管理 VM 的生命周期，是 OpenStack 中最核心的服务。

Neutron：为 OpenStack 提供网络连接服务，负责创建和管理 L2、L3 网络，为 VM 提供虚拟网络和物理网络连接。

Glance：管理 VM 启动镜像，Nova 创建 VM 时将使用 Glance 提供的镜像。

Cinder：为 VM 提供块存储服务。Cinder 提供的每一个 Volume 在 VM 看来就是一块虚拟硬盘，一般作为数据盘。

Swift：提供对象存储服务。VM 可以通过 RESTful API 存放对象数据。作为可选的方案，Glance 可以将镜像存放在 Swift 中；Cinder 也可以将 Volume 备份到 Swift 中。

Keystone：为 OpenStack 的各种服务提供认证和权限管理服务。简单地说，OpenStack 上的每一个操作都必须通过 Keystone 的审核。

Ceilometer：提供 OpenStack 监控和计量服务，为报警、统计或计费提供数据。

Horizon：为 OpenStack 用户提供一个 Web 的自服务图形化界面。

在上面这些服务中，哪些是 OpenStack 核心服务呢？核心服务是：如果没有它，OpenStack 就跑不起来。

很显然：Nova 管理计算资源，核心服务；Neutron 管理网络资源，核心服务；Glance 为云主机提供 OS 镜像，属于存储范畴，核心服务；Cinder 提供块存储，云主机需要数据盘，核心服务；Swift 提供对象存储，不是必须的，可选服务；Keystone 认证服务，为 OpenStack 其他服务提供认证，核心服务；Ceilometer 监控服务，不是必须的，可选服务；Horizon 为用户提供一个操作界面，不是必需的。

3.2 准备工作

3.2.1 OpenStack 环境部署

OpenStack 是一个分布式系统，由不同的组件共同支持整个云平台的运行，根据不同的需求，能够灵活地设计整个 OpenStack 架构，不同的组件装在不同的物理机上，甚至某个组件下的子服务也可装在不同的物理机上。一般来说，一个 OpenStack 平台由以下功能节点（Node）组成。

1. 控制节点（Controller Node）

管理 OpenStack，其上运行的服务有 Keystone、Glance、Horizon 以及 Nova 和 Neutron 中管理相关的组件。控制节点也运行支持 OpenStack 的服务，例如 SQL 数据库（通常是 MySQL）、消息队列（通常是 RabbitMQ）和网络时间服务 NTP。

2. 网络节点（Network Node）

其上运行的服务为 Neutron。为 OpenStack 提供 L2 和 L3 网络。包括虚拟机网络、DHCP、路由、NAT 等。

3. 存储节点（Storage Node）

提供块存储（Cinder）或对象存储（Swift）服务。

4. 计算节点（Compute Node）

其上运行 Hypervisor（默认使用 KVM）。同时运行 Neutron 服务的 Agent，为虚拟机提供网络支持。

这几类节点是从功能上进行的逻辑划分，在实际部署时可以根据需求灵活配置，比如：

（1）在大规模 OpenStack 生产环境中，每类节点都分别部署在若干台物理服务器上，各司其职并互相协作。这样的环境具备很好的性能、伸缩性和高可用性。

（2）在最小的实验环境中，可以将 4 类节点部署到一个物理的甚至是虚拟服务器上。麻雀虽小，五脏俱全，通常也称为 All-in-One 部署。

在我们的实验环境中，为了使得拓扑简洁同时功能完备，我们用两台计算机来部署 OpenStack Mitaka 版本。

物理机 1：控制节点，Centos7.2 1511（minal），1TB 硬盘，8GB 内存；

物理机 2：计算节点，Centos7.2 1511（minal），1TB 硬盘，8GB 内存。

注：控制节点同时作为控制节点和网络节点；

计算节点同时作为计算节点和存储节点。

本次部署架构图，如图 3-3 所示。

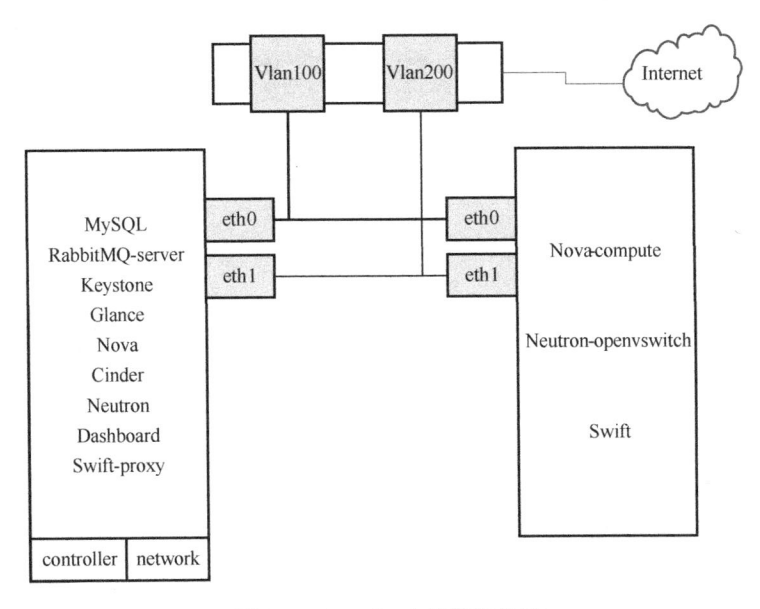

图 3-3　OpenStack 部署架构图

3.2.2　安全配置

1. 防火墙设置

CentOS 7 中默认启用了 Firewall 防火墙，在安装过程中，有些步骤可能会失败，除非你禁用或者修改防火墙规则。在入门学习中，我们将采用关闭防火墙的方法。

关闭防火墙：

 # systemctl mask firewalld.service
 # systemctl disable firewalld.service

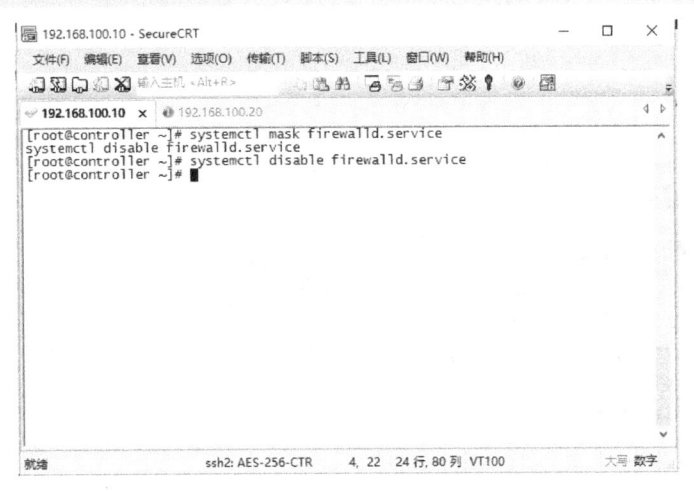

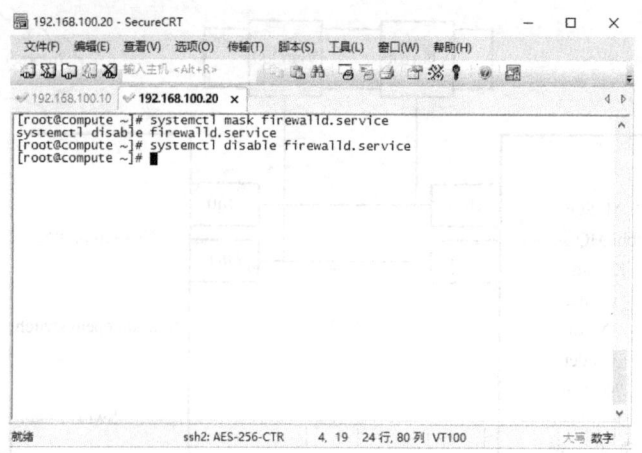

2. Selinux 设置

编辑 **vi /etc/selinux/config** 文件。

修改：

SELINUX=permissive/disabled

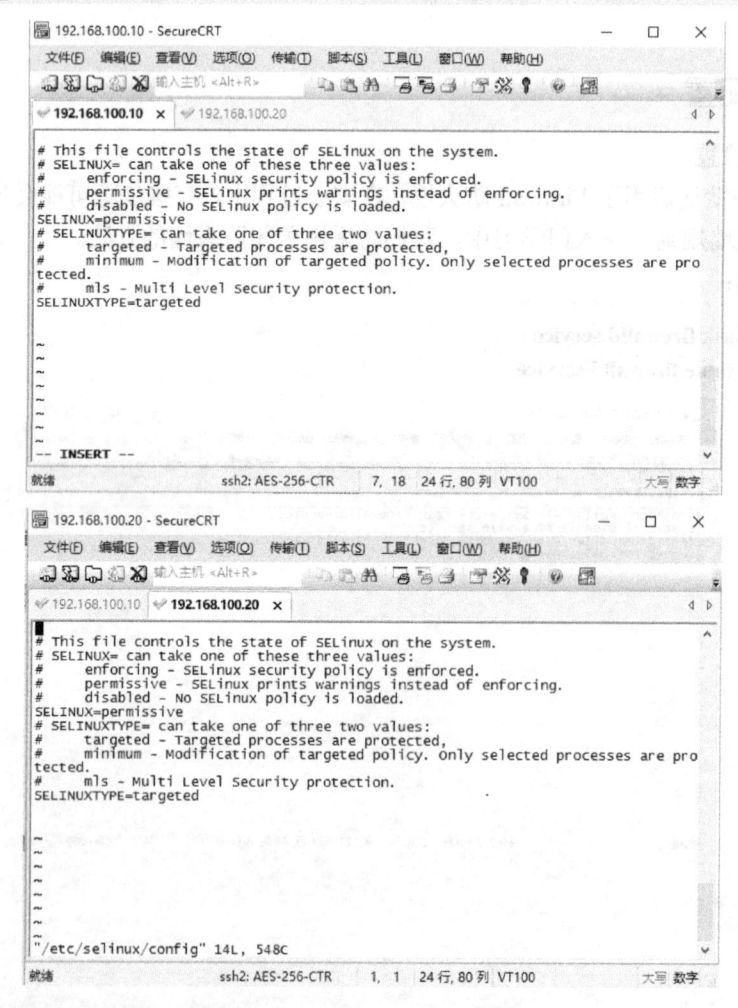

reboot 机器以生效防火墙配置或者 setenforce 0 生效配置

3.2.3 网络配置

网络的连通性非常重要，各个节点需要做到网络互相 Ping 通，使之处于同一个网络中。

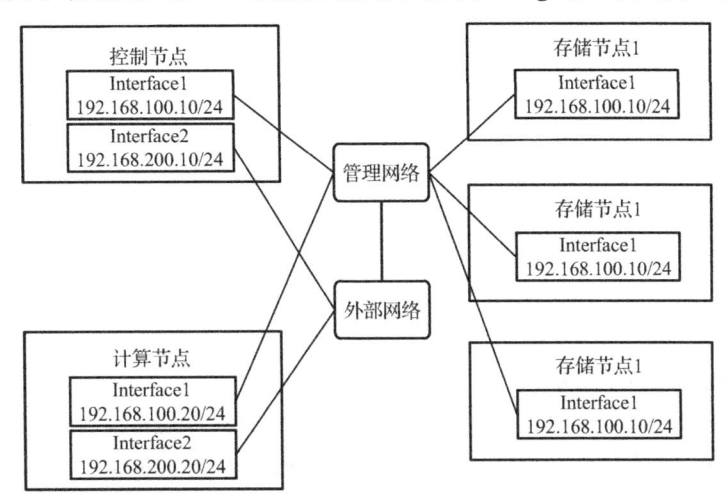

图 3-4 配置图

192.168.100 网段管理网络：这个网络为所有节点提供内部的管理目的的访问，例如包的安装、安全更新、DNS 和 NTP。

192.168.200 网段外部网络：实现虚拟机的访问。在外部网络中，所有虚拟机连接到外部网络。

配置 controller 节点网络信息。

网口 enp1s0f0：作为管理网络使用，配置 IP 为 **192.168.100.10/24**。

网口 enp1s0f1：作为外部网络使用，配置 IP 为 **192.168.200.10/24**。

vi /etc/sysconfig/network-scripts/ifcfg-enp1s0f0

```
TYPE=Ethernet
BOOTPROTO=static
DEFROUTE=yes
PEERDNS=yes
PEERROUTES=yes
IPV4_FAILURE_FATAL=no
IPV6INIT=yes
IPV6_AUTOCONF=yes
IPV6_DEFROUTE=yes
IPV6_PEERDNS=yes
IPV6_PEERROUTES=yes
IPV6_FAILURE_FATAL=no
NAME=enp1s0f1
UUID=f16bf581-8f8b-4e4c-a650-b82e330b4d0d
DEVICE=enp1s0f1
ONBOOT=yes
IPADDR=192.168.200.10
NETMASK=255.255.255.0
GATEWAY=192.168.200.1
```

vi /etc/sysconfig/network-scripts/ifcfg- enp1s0f1

```
TYPE=Ethernet
BOOTPROTO=static
DEFROUTE=yes
PEERDNS=yes
PEERROUTES=yes
IPV4_FAILURE_FATAL=no
IPV6INIT=yes
IPV6_AUTOCONF=yes
IPV6_DEFROUTE=yes
IPV6_PEERDNS=yes
IPV6_PEERROUTES=yes
IPV6_FAILURE_FATAL=no
NAME=enp1s0f0
UUID=f1a9f807-909c-4382-bdb8-03023c46fc64
DEVICE=enp1s0f0
ONBOOT=yes
IPADDR=192.168.100.10
NETMASK=255.255.255.0
GATEWAY=192.168.100.1
```

配置 Compute 节点内外网口。

网口 enp1s0f0：作为管理网络使用，配置 IP 为 **192.168.100.20/24**。

网口 enp1s0f1：作为外部网络使用，配置 IP 为 **192.168.200.20/24**。

vi /etc/sysconfig/network-scripts/ifcfg-enp1s0f0

```
TYPE=Ethernet
BOOTPROTO=static
DEFROUTE=yes
PEERDNS=yes
PEERROUTES=yes
IPV4_FAILURE_FATAL=no
IPV6INIT=yes
IPV6_AUTOCONF=yes
IPV6_DEFROUTE=yes
IPV6_PEERDNS=yes
IPV6_PEERROUTES=yes
IPV6_FAILURE_FATAL=no
NAME=enp1s0f0
UUID=4dcbc599-ff10-4743-a6cd-a2af18e60244
DEVICE=enp1s0f0
ONBOOT=yes
IPADDR=192.168.100.20
NETMASK=255.255.255.0
GATEWAY=192.168.100.1
```

vi /etc/sysconfig/network-scripts/ifcfg-enp1s0f1

```
TYPE=Ethernet
BOOTPROTO=static
DEFROUTE=yes
PEERDNS=yes
PEERROUTES=yes
IPV4_FAILURE_FATAL=no
IPV6INIT=yes
IPV6_AUTOCONF=yes
IPV6_DEFROUTE=yes
IPV6_PEERDNS=yes
IPV6_PEERROUTES=yes
IPV6_FAILURE_FATAL=no
NAME=enp1s0f1
UUID=9ba11e50-4a4f-4282-94c7-bb210bebc5c7
DEVICE=enp1s0f1
ONBOOT=yes
IPADDR=192.168.200.20
NETMASK=255.255.255.0
GATEWAY=192.168.200.1
```

重启网络检测连通性。

 # service network restart

3.2.4 配置主机映射

修改 **/etc/hosts** 文件添加以下内容。

192.168.100.10 controller
192.168.100.20 compute

1. 控制节点

2. 计算节点

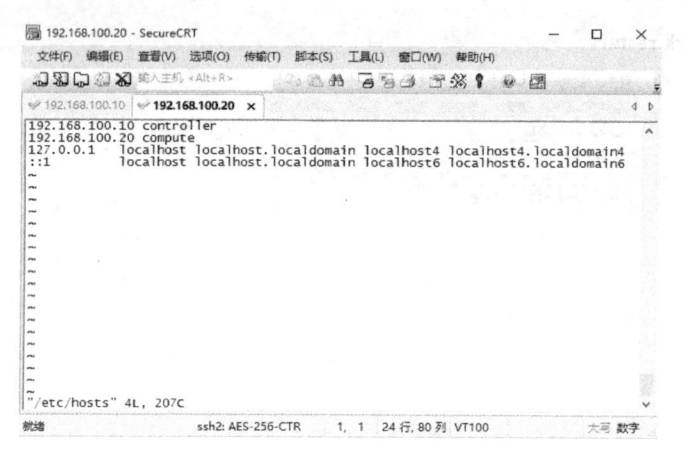

Ping 连通性测试：

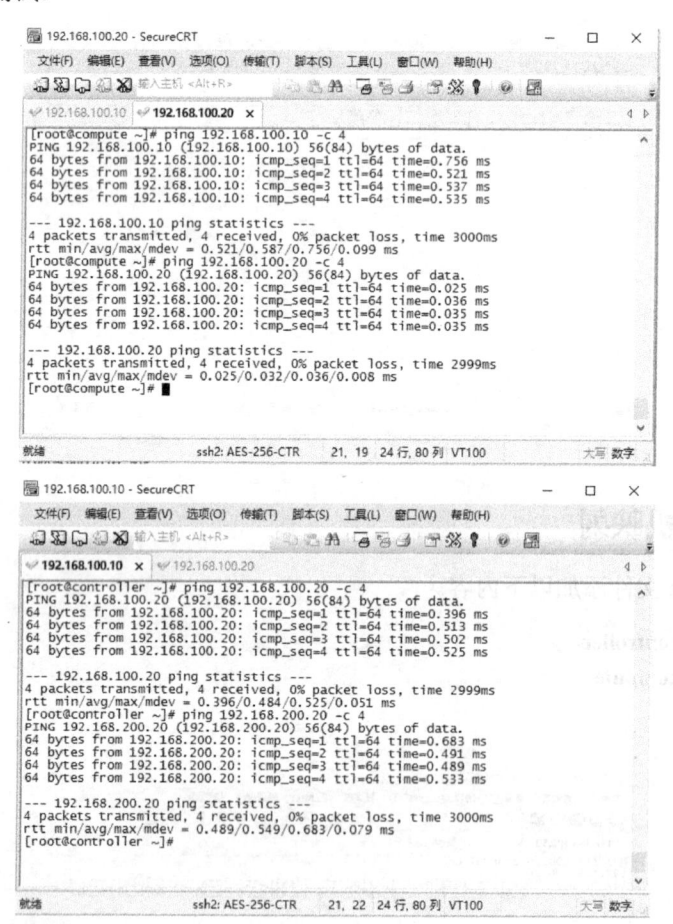

3.2.5 配置 yum 源

先使用 SecureFX 工具将所需软件包镜像文件上传至控制节点"/"目录下，然后进行挂载：

mount -o loop /CentOS-7-x86_64-DVD-1511.iso /mnt/centos
mount -o loop /Mitaka.iso /mnt/mitaka

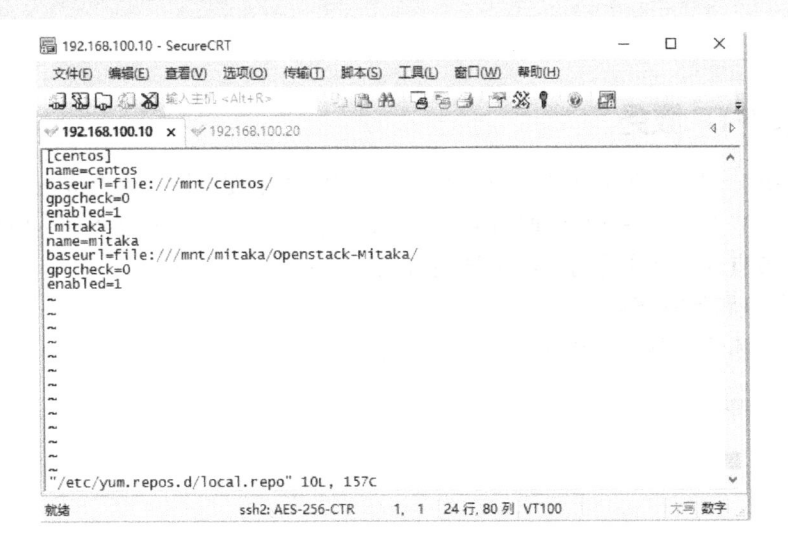

Yum 源备份

mv /etc/yum.repos.d/* /opt/

注：本节先创建 local.repo 文件再安装 FTP 服务。

配置 repo 文件

在/etc/yum.repo.d/下创建 local.repo 文件，搭建 FTP 服务器指向存放 yum 源文件路径。

控制节点安装 FTP 服务：

yum install -y vsftpd

添加配置项：

# vi /etc/vsftpd/vsftpd.conf
anon_root=/mnt

启动服务：

systemctl start vsftpd.service

注：注意防火墙是否处于关闭状态，源文件包存放于控制节点，在控制节点搭建 FTP 服务器。

1．控制节点

vi /etc/yum.repos.d/local.repo
[centos]
name=centos
baseurl=file:///mnt/centos/
（注：具体 yum 源根据真实环境配置）
gpgcheck=0
enabled=1
[mitaka]
name=mitaka
baseurl=file:///mnt/mitaka/Openstack-Mitaka/
（注：具体 yum 源根据真实环境配置）
gpgcheck=0
enabled=1

2. 计算节点

```
[centos]
name=centos
baseurl=ftp://192.168.100.10/centos/
（注：具体 yum 源根据真实环境配置）
gpgcheck=0
enabled=1
[mitaka]
name=mitaka
baseurl=ftp://192.168.100.10/mitaka/Openstack-Mitaka/
（注：具体 yum 源根据真实环境配置）
gpcheck=0
enabled=1
```

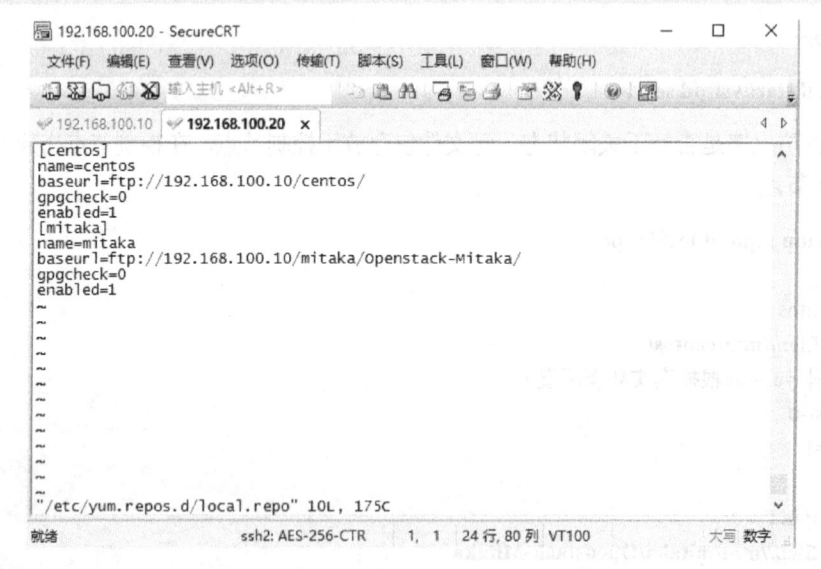

清除缓存：

yum clean all

注：两个节点都执行。

3.2.6 安装 NTP 服务

NTP 服务是一种时钟同步服务，在分布式集群中，为了便于同一生命周内不同节点服务的管理，需要各个节点的服务严格的时钟同步，在以下配置中，以控制节点作为时钟服务器，其他节点以控制节点的时钟作为时钟标准调整自己的时钟。

1. 控制节点和计算节点

在控制节点和计算节点安装 NTP 服务软件包。

yum install ntp -y

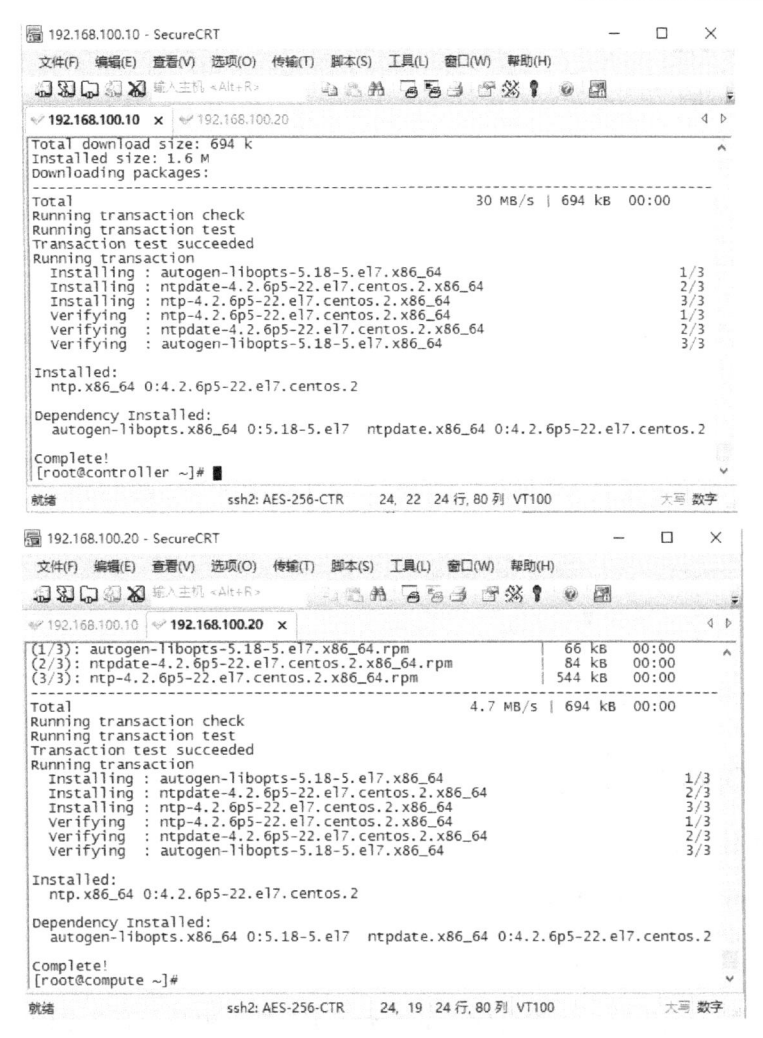

2. 配置控制节点

编辑 **/etc/ntp.conf** 文件。

添加以下内容：

server 127.127.1.0
fudge 127.127.1.0 stratum 10

启动 NTP 服务并设置开机自启：

 # systemctl start ntpd.service
 # systemctl enable ntpd.service
 # ntpstat

这一步是为了查看状态，直到显示第二次的结果即成功。

```
[root@controller ~]# ntpstat
synchronised to local net at stratum 6
   time correct to within 3948 ms
   polling server every 64 s
[root@controller ~]#
```

3. 配置计算节点

`# ntpdate controller`

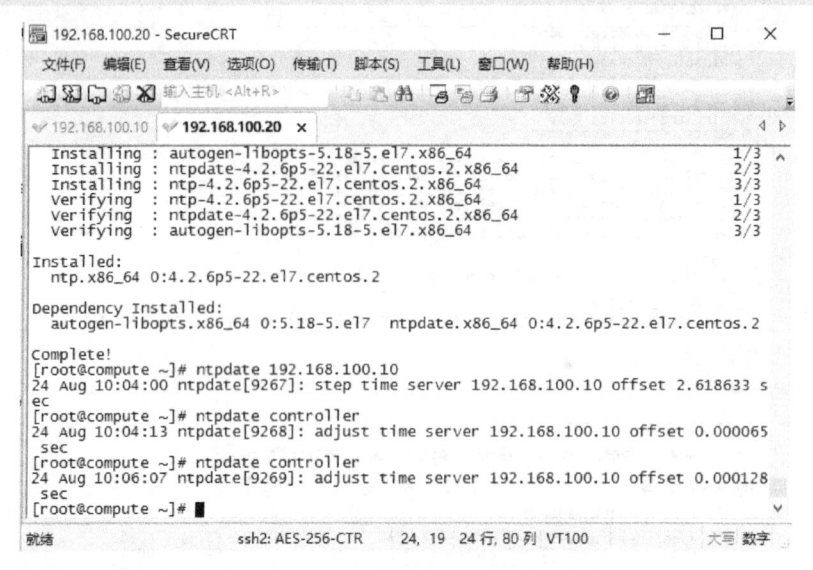

```
Installing  : autogen-libopts-5.18-5.el7.x86_64              1/3
Installing  : ntpdate-4.2.6p5-22.el7.centos.2.x86_64         2/3
Installing  : ntp-4.2.6p5-22.el7.centos.2.x86_64             3/3
Verifying   : ntp-4.2.6p5-22.el7.centos.2.x86_64             1/3
Verifying   : ntpdate-4.2.6p5-22.el7.centos.2.x86_64         2/3
Verifying   : autogen-libopts-5.18-5.el7.x86_64              3/3

Installed:
  ntp.x86_64 0:4.2.6p5-22.el7.centos.2

Dependency Installed:
  autogen-libopts.x86_64 0:5.18-5.el7    ntpdate.x86_64 0:4.2.6p5-22.el7.centos.2

Complete!
[root@compute ~]# ntpdate 192.168.100.10
24 Aug 10:04:00 ntpdate[9267]: step time server 192.168.100.10 offset 2.618633 sec
[root@compute ~]# ntpdate controller
24 Aug 10:04:13 ntpdate[9268]: adjust time server 192.168.100.10 offset 0.000065 sec
[root@compute ~]# ntpdate controller
24 Aug 10:06:07 ntpdate[9269]: adjust time server 192.168.100.10 offset 0.000128 sec
[root@compute ~]#
```

注：NTP 服务需要在每个节点上安装，并与控制节点同步。

3.2.7 安装 OpenStack 包

控制节点和计算节点

`# yum install python-openstackclient -y`
`# yum install openstack-selinux -y`
`# yum upgrade --skip-broken -y`

注：如果遇到安装失败，则检查 **/etc/yum.repo.d/** 文件夹中是否有多余的文件，若有，则删除。

3.2.8 安装并配置 SQL 数据库

SQL 数据库作为基础或扩展服务产生的数据存放的地方，数据库运行在控制节点上。OpenStack 支持的数据库有 MySQL、MariaDB 以及 PostgreSQL 等其他数据库。本次安装采用 MariaDB 数据库。

小助手：本次安装会大量地编辑配置文件，但是很多配置文件有许多以#开头的注释文件或者空格的命令，不容易找到自己需要编辑的模块。可采用下面的命令，删除#和空格的命令。

`# cat file |grep -v ^# |grep -v ^$ > newfile`

例如：修改 **/etc/my.cnf** 文件。

1. 备份 my.cnf 文件

`# cp /etc/my.cnf /etc/my.cnf.bak`

2. 删除#和空格的命令

cat /etc/my.cnf.bak |grep -v ^# |grep -v ^$ > /etc/my.cnf

安装数据库

数据库作为 OpenStack 的数据存储，OpenStack 支持多种数据库，比如 MariaDB-server，NoSQL 等。本次实验我们以 MariaDB 为例。

控制节点：

yum install mariadb mariadb-server python2-PyMySQL -y

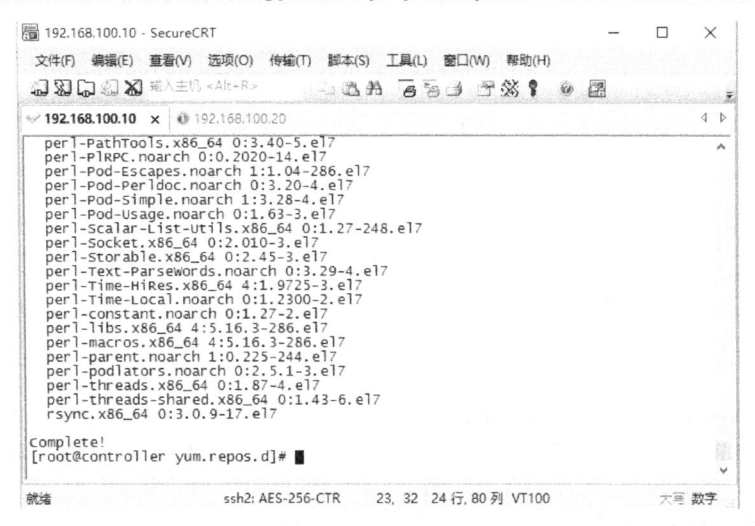

配置数据库

控制节点：

编辑 **/etc/my.cnf** 文件。

在**[mysqld]**部分配置 **bind-address** 值为控制节点的管理网络 IP 地址，以使得其他节点可以通过管理网络访问数据库。

[mysqld]
bind-address = 192.168.100.10

在**[mysqld]**部分配置如下键值来启用一些有用的选项和 UTF-8 字符集。

[mysqld]
default-storage-engine = innodb
innodb_file_per_table
max_connections = 4096
collation-server = utf8_general_ci
character-set-server = utf8

启动数据库：

systemctl enable mariadb.service
systemctl start mariadb.service

为了保证数据库服务的安全性，运行"**mysql_secure_installation**"脚本。初始化数据库并设置密码：

mysql_secure_installation

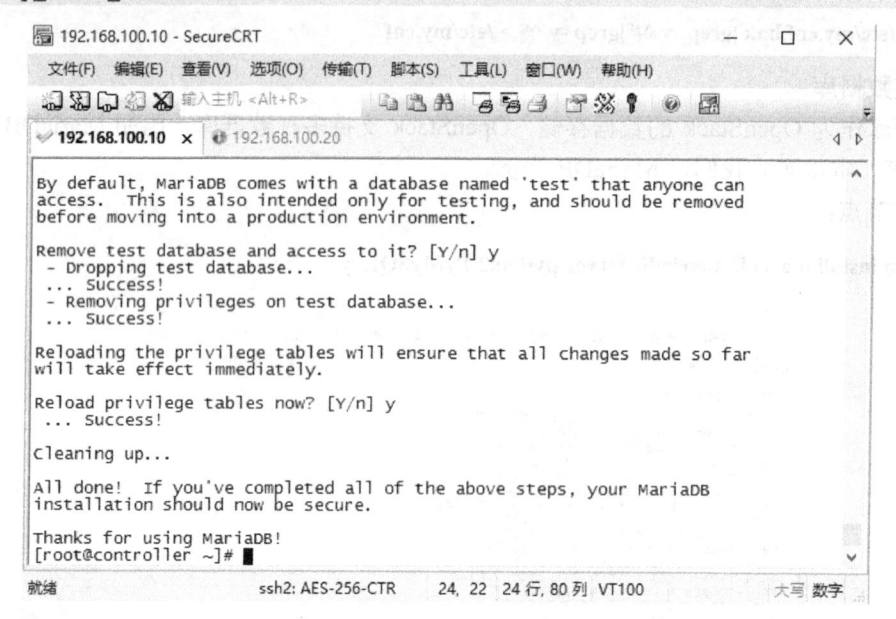

```
By default, MariaDB comes with a database named 'test' that anyone can
access.  This is also intended only for testing, and should be removed
before moving into a production environment.

Remove test database and access to it? [Y/n] y
 - Dropping test database...
 ... Success!
 - Removing privileges on test database...
 ... Success!

Reloading the privilege tables will ensure that all changes made so far
will take effect immediately.

Reload privilege tables now? [Y/n] y
 ... Success!

Cleaning up...

All done!  If you've completed all of the above steps, your MariaDB
installation should now be secure.

Thanks for using MariaDB!
[root@controller ~]#
```

初始化过程：

NOTE: RUNNING ALL PARTS OF THIS SCRIPT IS RECOMMENDED FOR ALL MariaDB
    SERVERS IN PRODUCTION USE!   PLEASE READ EACH STEP CAREFULLY!

In order to log into MariaDB to secure it, we'll need the current
password for the root user.   If you've just installed MariaDB, and
you haven't set the root password yet, the password will be blank,
so you should just press enter here.

Enter current password for root (enter for none): (第一次输入为回车，因为没有密码)
OK, successfully used password, moving on...

Setting the root password ensures that nobody can log into the MySQL
root user without the proper authorisation.

Set root password? [Y/n] y(第二次输入为 y，然后设置数据库密码)
New password:
Re-enter new password:
Password updated successfully!
Reloading privilege tables..
    ... Success!

By default, a MySQL installation has an anonymous user, allowing anyone
to log into MySQL without having to have a user account created for
them. This is intended only for testing, and to make the installation
go a bit smoother.   You should remove them before moving into a
production environment.

```
Remove anonymous users? [Y/n] y(第三次输入为 y)
 ... Success!

Normally, root should only be allowed to connect from 'localhost'.   This
ensures that someone cannot guess at the root password from the network.

Disallow root login remotely? [Y/n] n(第四次输入为 n)
 ... skipping.

By default, MySQL comes with a database named 'test' that anyone can
access.   This is also intended only for testing, and should be removed
before moving into a production environment.

Remove test database and access to it? [Y/n] y(第五次输入为 y)
 - Dropping test database...
 ... Success!
 - Removing privileges on test database...
 ... Success!

Reloading the privilege tables will ensure that all changes made so far
will take effect immediately.

Reload privilege tables now? [Y/n] y(第六次输入为 y)
 ... Success!

Cleaning up...

All done!   If you've completed all of the above steps, your MySQL
installation should now be secure.

Thanks for using MySQL!
```

3.2.9 安装并配置消息服务器

安装并启动消息服务器

OpenStack 使用 message queue 协调操作和各服务的状态信息。消息队列服务本次部署在控制节点上。OpenStack 支持的几种消息队列服务包括 RabbitMQ、Qpid 和 ZeroMQ。我们采用安装 RabbitMQ 消息队列服务。

控制节点：

```
# yum install rabbitmq-server -y
```

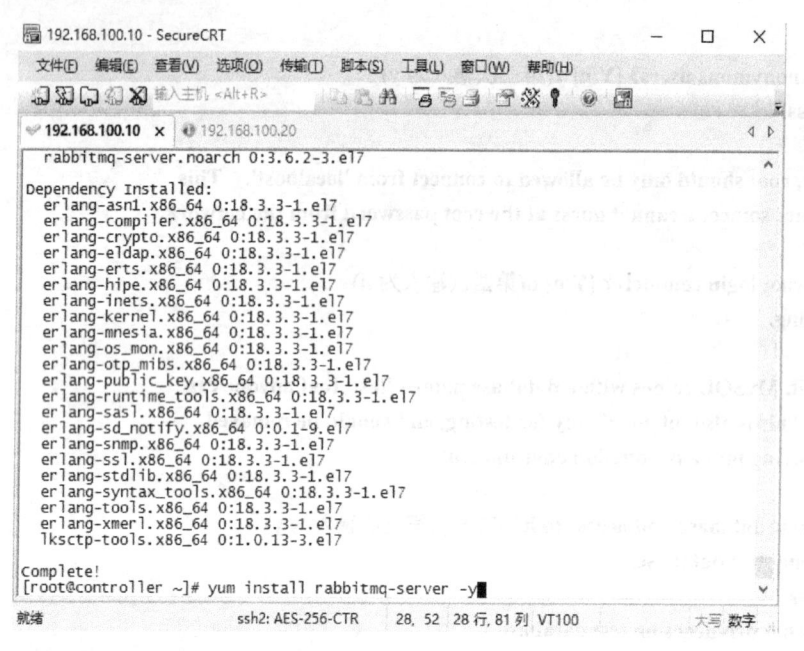

启动 RabbitMQ 服务并设置开机自启动：

 # systemctl enable rabbitmq-server.service
 # systemctl start rabbitmq-server.service

创建 rabbitmq 用户并设置权限
创建用户：**openstack**；密码：**000000**。

 # rabbitmqctl add_user openstack 000000

给 OpenStack 用户授予读/写访问权限。

 # rabbitmqctl set_permissions openstack ".*" ".*" ".*"

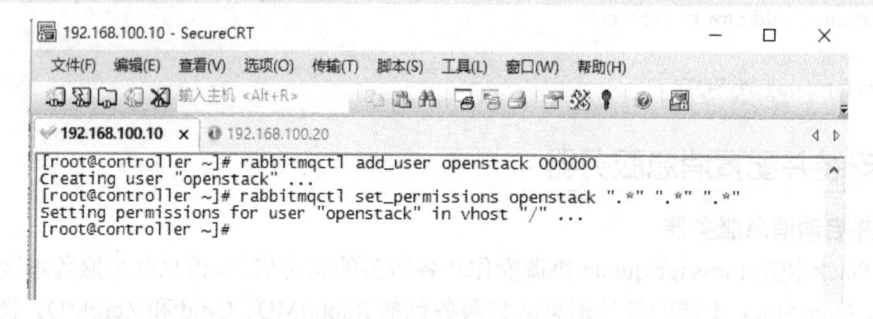

3.2.10 安装 Memcached

认证服务认证缓存使用 Memcached 缓存令牌，缓存服务 memecached 运行在控制节点上。在生产部署中，建议联合启用防火墙、认证和加密保证它的安全。
控制节点：

 # yum install memcached python-memcached -y

启动 Memcached 服务并设置开机自启动：

systemctl enable memcached.service
systemctl start memcached.service

课后习题

1. 以下哪个是 OpenStack 中最核心的服务？（　　）
A．Node　　　　　　B．Nova　　　　　　C．Neutron　　　　　　D．VM
2. 以下哪个不是 OpenStack 中的核心服务？（　　）
A．Neutron　　　　　B．Keystone　　　　C．Swift　　　　　　　D．nova
3. Keystone 为 OpenStack 的各种服务提供_____、_____和服务。
4. Ceilometer 为 OpenStack 提供了_____、_____和服务。
5. 为 OpenStack 提供对象存储服务的组件是_____。
6. 控制节点上运行的 OpenStack 组件有_____。控制节点也运行支持 OpenStack 的服务，例如_____。
7. 请写出 OpenStack 的核心服务并选择其中两个服务并简述它的作用。

第 4 章

认证服务 Keystone

4.1 Keystone 基本概念

Keystone 是 OpenStack 的身份认证服务,当安装 OpenStack 身份认证服务时,必须将之注册到其 OpenStack 安装环境的每个服务,身份认证服务才可以追踪那些已经安装的 OpenStack 服务,以及在网络中定位它们。Keystone 组成主要分为以下部分。

域(**Domain**):Domain 实现真正的多租户(multi-tenancy)架构,Domain 担任 Project 的高层容器。云服务的客户是 Domain 的所有者,他们可以在自己的 Domain 中创建多个 Projects、Users、Groups 和 Roles。通过引入 Domain,云服务客户可以对其拥有的多个 Project 进行统一管理,而不必再像过去那样对每一个 Project 进行单独管理。

用户(**User**):那些使用 OpenStack 云服务的人、系统、服务的数字表示。身份认证服务会验证那些生成调用的用户发过来的请求,用户登录且被赋予令牌以访问资源,用户可以直接被分配到特别的租户和行为,如果他们是被包含在租户中的。

凭证(**Credential**):用户身份的确认数据,例如,用户名和密码、用户名和 API 密钥,或者是一个由身份服务提供的授权令牌。

认证(**Authentication**):确认用户身份的流程,OpenStack 身份认证服务确认发过来的请求,即验证由用户提供的凭证。

令牌(**Token**):一个字母数字混合的文本字符串,用户访问 OpenStack API 和资源,令牌可以随时撤销,以及有一定的时间期限。

租户(**Project**):用于组成或隔离资源的容器,租户会组成或隔离身份对象,一个租户会映射到一个客户、一个账户、一个组织或一个项目。

服务(**Service**):一个 OpenStack 服务,如计算服务(Nova),对象服务(Swift),或镜像服务(Glance)。它提供一个或多个端点来让用户可以访问资源和执行操作。

角色(**Role**):定义了一组用户权限的用户,可赋予其执行某些特定的操作。在身份服务中，一个令牌所携带用户信息包含角色列表。服务在被调用时会看用户是什么样的角色，这个角色赋予的权限能够操作什么资源。

Keystone 客户端：为 OpenStack 身份 API 提供的一组命令行接口。例如，用户可以运行 keystone service-create 和 keystone endpoint-create 命令在其 OpenStack 环境中去注册服务。

策略（Policy）：OpenStack 对用户的验证除了 OpenStack 的身份验证以外，还需要鉴别用户对某个服务是否有访问权限。Policy 机制就是用来控制某一个 User 在某个 Tenant 中某个操作的权限。这个 User 能执行什么操作，不能执行什么操作，就是通过 Policy 机制来实现的。对于 Keystone 服务来说，Policy 就是一个 json 文件，通过配置这个文件（/etc/keystone/policy.json），Keystone Service 实现了对 User 的基于用户角色的权限管理。

端点（Endpoint）：一个网络可访问的服务地址，通过它可以访问一个服务，通常是个 URL 地址。不同 Region 有不同的 Service Endpoint。Endpoint 告诉 OpenStack Service 去哪里访问特定的 Servcie。比如，当 Nova 需要访问 Glance 服务去获取 Image 时，Nova 通过访问 Keystone 拿到 Glance 的 Endpoint，然后通过访问该 Endpoint 去获取 Glance 服务。我们可以通过 Endpoint 的 Region 属性去定义多个 Region。Endpoint 的使用对象分为以下三类。

Adminurl：给 admin 用户使用。

Internalurl：供 OpenStack 内部服务使用，以便与其他服务进行通信。

Publicurl：其他用户可以访问的地址。

4.2 Keystone 数据库操作

登录 MySQL 并创建 Keystone 数据库：

 # mysql -uroot -p000000

创建 Keystone 数据库：

MariaDB [(none)]>**CREATE DATABASE keystone;**

设置授权用户和密码：

MariaDB [(none)]>**GRANT ALL PRIVILEGES ON keystone.* TO 'keystone'@'%' IDENTIFIED BY '000000';**

MariaDB [(none)]> **GRANT ALL PRIVILEGES ON keystone.* TO 'keystone'@'localhost' IDENTIFIED BY '000000';**

MariaDB [(none)]>**exit**

4.3 安装并配置 Keystone

安装 Keystone 所需软件包：

 # yum install openstack-keystone httpd mod_wsgi -y

生成一个随机值作为初始配置期间的管理令牌：

openssl rand -hex 10

编辑 **/etc/keystone/keystone.conf** 文件。

做如下配置与修改。

使用刚刚生成的随机值替换：

[DEFAULT]
admin_token = 4d41e6e909f346df2676

配置数据库链接：

[database]
connection = mysql+pymysql://keystone:000000@controller/keystone

配置 provider：

 [token]
provider = fernet

同步数据库：

su -s /bin/sh -c "keystone-manage db_sync" keystone

注：进入 Keystone 数据库查看是否有数据表，验证是否同步成功。

初始化密钥：

keystone-manage fernet_setup --keystone-user keystone --keystone-group keystone

4.4 配置 Apache 服务

编辑 **/etc/httpd/conf/httpd.conf** 文件。

添加：

ServerName controller

创建 **/etc/httpd/conf.d/wsgi-keystone.conf** 文件。

添加：

Listen 5000
Listen 35357

<VirtualHost *:5000>
　　WSGIDaemonProcess keystone-public processes=5 threads=1 user=keystone group=keystone display-name=%{GROUP}
　　WSGIProcessGroup keystone-public
　　WSGIScriptAlias / /usr/bin/keystone-wsgi-public
　　WSGIApplicationGroup %{GLOBAL}
　　WSGIPassAuthorization On
　　ErrorLogFormat "%{cu}t %M"
　　ErrorLog /var/log/httpd/keystone-error.log
　　CustomLog /var/log/httpd/keystone-access.log combined

　　<Directory /usr/bin>
　　　　Require all granted
　　</Directory>
</VirtualHost>

<VirtualHost *:35357>
　　WSGIDaemonProcess keystone-admin processes=5 threads=1 user=keystone group=keystone display-name=%{GROUP}
　　WSGIProcessGroup keystone-admin
　　WSGIScriptAlias / /usr/bin/keystone-wsgi-admin
　　WSGIApplicationGroup %{GLOBAL}
　　WSGIPassAuthorization On
　　ErrorLogFormat "%{cu}t %M"
　　ErrorLog /var/log/httpd/keystone-error.log
　　CustomLog /var/log/httpd/keystone-access.log combined

　　<Directory /usr/bin>
　　　　Require all granted
　　</Directory>
</VirtualHost>

启动 Apache HTTP 服务并设置开机自启动：

systemctl enable httpd.service

systemctl start httpd.service

4.5 创建 Service 和 API Endpoints

1. 配置身份认证令牌

export OS_TOKEN=4d41e6e909f346df2676

2. 配置端点 URL

export OS_URL=http://controller:35357/v3

3. 配置 API 版本

export OS_IDENTITY_API_VERSION=3

4. 为 Keystone 本身创建服务

openstack service create --name keystone --description "OpenStack Identity" identity

```
[root@controller ~]# openstack service create \
> --name keystone --description "OpenStack Identity" identity
+-------------+----------------------------------+
| Field       | Value                            |
+-------------+----------------------------------+
| description | OpenStack Identity               |
| enabled     | True                             |
| id          | 7e8f30e8aff94cabbc69d61580d23484 |
| name        | keystone                         |
| type        | identity                         |
+-------------+----------------------------------+
```

然后，创建 Keystone 身份认证服务的端点：

身份认证服务管理了一个与环境相关的 API 端点目录。使用这个目录来决定如何与创建环境中的其他服务进行通信。

OpenStack 使用三个 API 端点代表每种服务：**admin**、**internal** 和 **public**。默认情况下，管理 API 端点允许修改用户和租户，而公共和内部 API 不允许这些操作。此次安装为所有端点和默认"RegionOne"区域都使用管理网络。

1. 创建公共端点

openstack endpoint create --region RegionOne identity public http://controller:5000/v3

2. 创建外部端点

openstack endpoint create --region RegionOne identity internal http://controller:5000/v3

3. 创建管理端点

openstack endpoint create --region RegionOne identity admin http://controller:35357/v3

```
[root@controller ~]# openstack endpoint create --region RegionOne \
>    identity public http://controller:5000/v3
+--------------+----------------------------------+
| Field        | Value                            |
+--------------+----------------------------------+
| enabled      | True                             |
| id           | f60740a1b90443ceb1bdce81af7dad93 |
| interface    | public                           |
| region       | RegionOne                        |
| region_id    | RegionOne                        |
| service_id   | 7e8f30e8aff94cabbc69d61580d23484 |
| service_name | keystone                         |
| service_type | identity                         |
| url          | http://controller:5000/v3        |
+--------------+----------------------------------+
[root@controller ~]# openstack endpoint create --region RegionOne \
>    identity internal http://controller:5000/v3
+--------------+----------------------------------+
| Field        | Value                            |
+--------------+----------------------------------+
| enabled      | True                             |
| id           | 1fb7fdf67f5b4d1daa5785d0105ca32c |
| interface    | internal                         |
| region       | RegionOne                        |
| region_id    | RegionOne                        |
| service_id   | 7e8f30e8aff94cabbc69d61580d23484 |
| service_name | keystone                         |
| service_type | identity                         |
| url          | http://controller:5000/v3        |
+--------------+----------------------------------+
[root@controller ~]# openstack endpoint create --region RegionOne \
>    identity admin http://controller:35357/v3
+--------------+----------------------------------+
| Field        | Value                            |
+--------------+----------------------------------+
| enabled      | True                             |
| id           | 8202f7ef9c184bc1beb9b8b20f1d56df |
| interface    | admin                            |
| region       | RegionOne                        |
| region_id    | RegionOne                        |
| service_id   | 7e8f30e8aff94cabbc69d61580d23484 |
| service_name | keystone                         |
| service_type | identity                         |
| url          | http://controller:35357/v3       |
+--------------+----------------------------------+
```

4.6 创建 domain、project、user、role

1. 创建默认（default）的 domain

openstack domain create --description "Default Domain" default

```
[root@controller ~]# openstack domain create --description "Default Domain" default
+-------------+----------------------------------+
| Field       | Value                            |
+-------------+----------------------------------+
| description | Default Domain                   |
| enabled     | True                             |
| id          | d63fbba811b94cffb2cf9f88b59f4066 |
| name        | default                          |
+-------------+----------------------------------+
```

2. 创建名字为 admin 的 project

openstack project create --domain default --description "Admin Project" admin

3. 创建名字为 admin 的 user

openstack user create --domain default --password-prompt admin（回车之后输入自定义密码）

```
[root@controller ~]# openstack project create --domain default \
>    --description "Admin Project" admin
+-------------+----------------------------------+
| Field       | Value                            |
+-------------+----------------------------------+
| description | Admin Project                    |
| domain_id   | d63fbba811b94cffb2cf9f88b59f4066 |
| enabled     | True                             |
| id          | e71744580698483c90b1ac0a03d66313 |
| is_domain   | False                            |
| name        | admin                            |
| parent_id   | d63fbba811b94cffb2cf9f88b59f4066 |
+-------------+----------------------------------+
[root@controller ~]# openstack user create --domain default \
>    --password-prompt admin
User Password:
Repeat User Password:
+-----------+----------------------------------+
| Field     | Value                            |
+-----------+----------------------------------+
| domain_id | d63fbba811b94cffb2cf9f88b59f4066 |
| enabled   | True                             |
| id        | d5c9272aadc24c00899fe2fe00f3aa7f |
| name      | admin                            |
+-----------+----------------------------------+
[root@controller ~]# openstack role create admin
+-----------+----------------------------------+
| Field     | Value                            |
+-----------+----------------------------------+
| domain_id | None                             |
| id        | 254853f3e6ba47f6aad28bd15e17c39b |
| name      | admin                            |
+-----------+----------------------------------+
[root@controller ~]# openstack role add --project admin --user admin admin
```

4. 创建名字为 admin 的 role

openstack role create admin

5. 进行关联

openstack role add --project admin --user admin admin

6. 创建名为 service 的 project

openstack project create --domain default --description "Service Project" service

7. 创建名为 demo 的 project

openstack project create --domain default --description "Demo Project" demo

8. 创建名为 demo 的 user

openstack user create --domain default --password-prompt demo（回车之后输入自定义密码）

9. 创建名为 demo 的 role

openstack role create user

10. 进行关联

openstack role add --project demo --user demo user

```
[root@controller ~]# openstack project create --domain default \
>    --description "Service Project" service
+-------------+----------------------------------+
| Field       | Value                            |
+-------------+----------------------------------+
| description | Service Project                  |
| domain_id   | d63fbba811b94cffb2cf9f88b59f4066 |
| enabled     | True                             |
| id          | b4817f0408b74c82948f57bada659aa7 |
| is_domain   | False                            |
| name        | service                          |
| parent_id   | d63fbba811b94cffb2cf9f88b59f4066 |
+-------------+----------------------------------+
[root@controller ~]# openstack project create --domain default \
>    --description "Demo Project" demo
+-------------+----------------------------------+
| Field       | Value                            |
+-------------+----------------------------------+
| description | Demo Project                     |
| domain_id   | d63fbba811b94cffb2cf9f88b59f4066 |
| enabled     | True                             |
| id          | 622d72c12b9b4fb49b76860b3178f490 |
| is_domain   | False                            |
| name        | demo                             |
| parent_id   | d63fbba811b94cffb2cf9f88b59f4066 |
+-------------+----------------------------------+
[root@controller ~]# openstack user create --domain default \
>    --password-prompt demo
User Password:
Repeat User Password:
+-----------+----------------------------------+
| Field     | Value                            |
+-----------+----------------------------------+
| domain_id | d63fbba811b94cffb2cf9f88b59f4066 |
| enabled   | True                             |
| id        | 03a2b22588e74452ae3b2c1b803c1e6a |
| name      | demo                             |
+-----------+----------------------------------+
[root@controller ~]# openstack role create user
+-----------+----------------------------------+
| Field     | Value                            |
+-----------+----------------------------------+
| domain_id | None                             |
| id        | eaa412db64644e37b1a075db189b8bea |
| name      | user                             |
+-----------+----------------------------------+
[root@controller ~]# _openstack role add --project demo --user demo user
```

4.7 验证 Keystone 服务

1. 不生效临时的环境变量

 # unset OS_TOKEN OS_URL

2. 查看 admin 用户，请求身份验证令牌

 # openstack --os-auth-url http://controller:35357/v3 --os-project-domain-name default --os-user-domain-name default --os-project-name admin --os-username admin tokenissue

```
[root@controller ~]# openstack --os-auth-url http://controller:35
357/v3  --os-project-domain-name default --os-user-domain-name d
efault  --os-project-name admin --os-username admin tokenissue
openstack: 'tokenissue' is not an openstack command. See 'opensta
ck --help'.
Did you mean one of these?
  extension list
```

3. 查看 demo 用户，请求身份验证令牌

openstack --os-auth-url http://controller:35357/v3 --os-project-domain-name default --os-user- domain-name default --os-project-name demo --os-username demo token issue

```
[root@controller ~]# openstack --os-auth-url http://controller:35357/v3 --os-project-domain-name default --os-user-domain-name default --os-project-name admin --os-username admin token issue
Password:
+------------+----------------------------------------------------------+
| Field      | Value                                                    |
+------------+----------------------------------------------------------+
| expires    | 2016-08-24T18:39:29.778816Z                              |
| id         | gAAAAABXvdvRHfFtSKV8dSrN1qqUCzqosOH_yCMHuZb-LtUS9z51mz7YfXx40Un |
|            | Gedpp9GCxH5h4DTP24Cele0DNrSTfWsUwyFT3qNhOnUrGNRK-DsChAbzTk8eRDt |
|            | iP2O5vQucLMQBXrheHmrzMN7RGANYFgksAUbY8FbBRd7O_eyC9cSpnlrQ |
| project_id | e71744580698483c90b1ac0a03d66313                         |
| user_id    | d5c9272aadc24c00899fe2fe00f3aa7f                         |
+------------+----------------------------------------------------------+
[root@controller ~]# openstack --os-auth-url http://controller:35357/v3 --os-project-domain-name default --os-user-domain-name default --os-project-name demo --os-username demo token issue
Password:
+------------+----------------------------------------------------------+
| Field      | Value                                                    |
+------------+----------------------------------------------------------+
| expires    | 2016-08-24T18:40:07.995144Z                              |
| id         | gAAAAABXvdv4mcjArjrKQ8782tD7YsF6domU7qdfXUlEjnQtT37f6VC2yI_h1nw |
|            | r_UKduvuDxYaP4wtJeIbcNs4jdepprmjjoG6-_SvNYb5Qy0 |
|            | -45KrkUt0Xhl2uKgitupAydMvBpD9MRwdazWuZgpX- |
|            | hVMFGbcs_gcUwIm_h7ikxn_oz8PptEg                          |
| project_id | 622d72c12b9b4fb49b76860b3178f490                         |
| user_id    | 03a2b22588e74452ae3b2c1b803c1e6a                         |
+------------+----------------------------------------------------------+
```

注：回车输入之间设置的用户的密码。

4. 写入环境变量

创建 **/root/admin-openrc** 文件。

添加：

export OS_PROJECT_DOMAIN_NAME=default
export OS_USER_DOMAIN_NAME=default
export OS_PROJECT_NAME=admin
export OS_USERNAME=admin
export OS_PASSWORD=000000
export OS_AUTH_URL=http://controller:35357/v3
export OS_IDENTITY_API_VERSION=3
export OS_IMAGE_API_VERSION=2

创建 **/root/demo-openrc** 文件。

添加：

export OS_PROJECT_DOMAIN_NAME=default
export OS_USER_DOMAIN_NAME=default
export OS_PROJECT_NAME=demo
export OS_USERNAME=demo
export OS_PASSWORD=000000
export OS_AUTH_URL=http://controller:5000/v3
export OS_IDENTITY_API_VERSION=3
export OS_IMAGE_API_VERSION=2

5. 生效并验证

\# . admin-openrc
\# openstack token issue

```
[root@controller ~]# . admin-openrc
[root@controller ~]# openstack token issue
+------------+-------------------------------------------------------------+
| Field      | value                                                       |
+------------+-------------------------------------------------------------+
| expires    | 2016-08-24T18:48:45.404171Z                                 |
| id         | gAAAAABXvd39Sap_uFDOiqQyillMFKTzCFerX1vVimtcSE9Vr4QzhdltNOK_TJx |
|            | vebaSH-lwUQoJ88H9JA-ifbIwQoajP7q7opcA1XYAqqfgv6m-yczyKqeZETVCJQ |
|            | KCOfBaSqIux3JUYRetTMI8gnIIXN3_Hg5bDmq_bibzU4XoMcooh9waQ60    |
| project_id | e71744580698483c90b1ac0a03d66313                            |
| user_id    | d5c9272aadc24c00899fe2fe00f3aa7f                            |
+------------+-------------------------------------------------------------+
```

Keystone 组件用户（User）与镜像服务（Glance）、计算服务（Nova）、网络服务（Neutron）等其他服务的关系如图 4-1 所示。

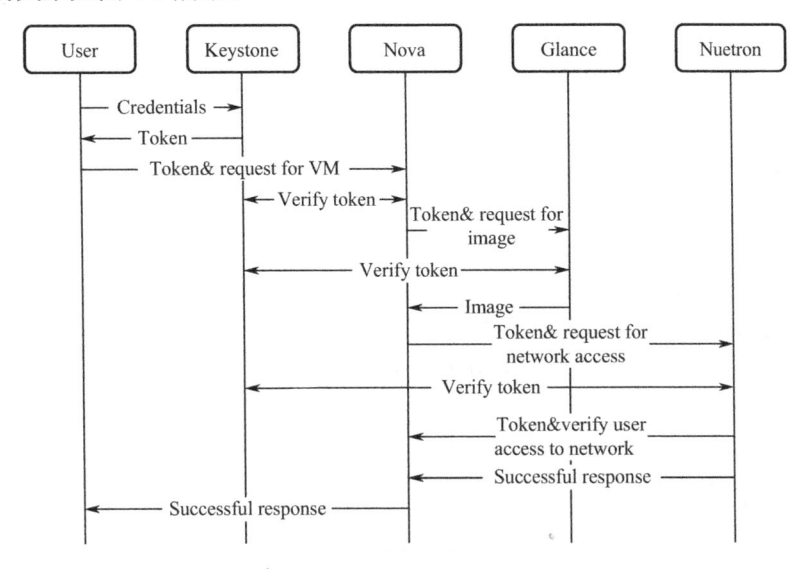

图 4-1 Keystone 组件用户与其他服务的关系

Keystone 与其他 OpenStack Service 之间的交互和协同工作：首先 User 向 Keystone 提供自己的 Credentials（凭证：用于确认用户身份的数据）。Keystone 会从 SQL Database 中读取数据对 User 提供的 Credentials 进行验证，如验证通过，会向 User 返回一个 Token，该 Token 限定了可以在有效时间内被访问的 OpenStack API Endpoint 和资源。此后，User 所有的 Request 都会使用该 Token 进行身份验证。如用户向 Nova 申请虚拟机服务，Nova 会将 User 提供的 Token 发送给 Keystone 进行 Verify 验证，Keystone 会根据 Token 判断 User 是否拥有执行申请虚拟机操作的权限，若验证通过，则 Nova 会向其提供相对应的服务。其他 Openstack 组件和 Keystone 的交互也是如此。

所以，Keystone 在整个 OpenStack 组件中的位置如图 4-2 所示。

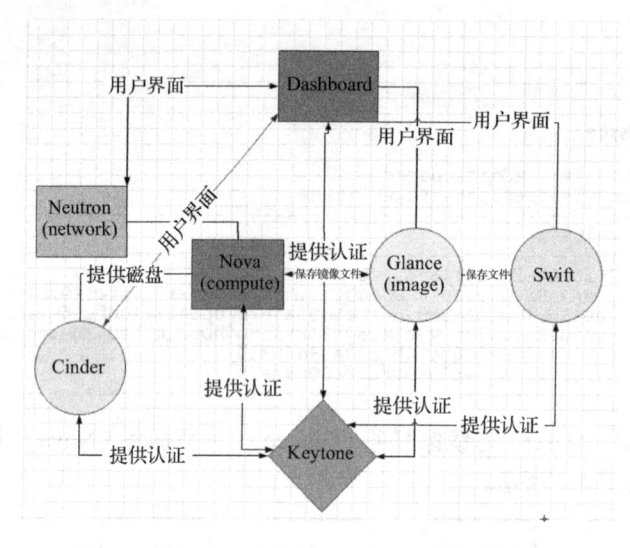

图 4-2 Keystone 在整个 OpenStack 组件中的位置

课后习题

1．Keystone 是 OpenStack 的_____服务，当安装 OpenStack 的_____服务，必须将之注册到其 OpenStack 安装环境的每个服务。身份服务才可以追踪哪些 OpenStack 服务已经安装，以及在网络中定位它们。其组成主要分为以下几部分：_____、_____、_____、_____、_____、_____、_____、_____、_____、_____。

2．同步 Keystone 数据库的命令是_____。

3．生成随机值的命令是_____。

4．Keystone 服务管理端点的端口号是_____。

5．如何创建一个名为 cqcet 的 project？

6．如何用命令创建一个名为 cqcet 的数据库，并设置授权 cqcet 用户和密码（密码为自己学号）？

第5章 镜像服务 Glance

5.1 Glance 基本概念

Glance 是 OpenStack 镜像服务，用来注册、登录和检索虚拟机镜像。Glance 服务提供了一个 REST API，使用户能够查询虚拟机镜像元数据和检索的实际镜像。通过镜像服务提供的虚拟机镜像可以存储在不同的位置，从简单的文件系统对象存储到类似 OpenStack 对象存储系统。

为简单起见，本次安装镜像服务使用普通文件系统作为存储后端，也就是说上传镜像将被存储在一个目录里，这个目录是控制节点的一个目录，但要确保这个目录提供足够的空间，然后再存储虚拟机的镜像和快照。

OpenStack 镜像服务包含以下两个组件。

（1）Glance-api：接收镜像 API 的调用，如镜像发现、恢复、存储。

（2）Glance-registry：存储、处理和恢复镜像的元数据，元数据包括大小和类型等项。registry 是一个私有的内部服务，这意味着仅为 OpenStack 镜像服务所使用。不要将它开放给用户数据库存放镜像元数据，用户是可以依据个人喜好选择数据库的，多数的部署使用 MySQL 或 SQLite。镜像文件的存储仓库支持多种类型的仓库，它们有普通文件系统、对象存储、RADOS 块设备、HTTP、亚马逊 S3。但是，其中一些仓库仅支持只读方式使用。

5.2 数据库配置

登录 MySQL 并创建 Glance 数据库：

`# mysql -uroot -p000000`

创建 Glance 数据库：

MariaDB [(none)]> **CREATE DATABASE glance;**

设置授权用户和密码：

MariaDB [(none)]>**GRANT ALL PRIVILEGES ON glance.* TO 'glance'@'%' IDENTIFIED BY '000000';**

MariaDB [(none)]> **GRANT ALL PRIVILEGES ON glance.* TO 'glance'@'localhost' IDENTIFIED BY '000000';**

MariaDB [(none)]>**exit**

5.3 创建服务凭证和 API 端点

1. 生效 admin 用户环境变量

 # . admin-openrc

2. 创建服务凭证

创建名为 glance 的用户（user）：

 # openstack user create --domain default --password-prompt glance

进行关联：添加 admin 角色到 glance 用户和 service 租户。

 # openstack role add --project service --user glance admin

创建 Glance 服务实体认证：

 # openstack service create --name glance --description "OpenStack Image" image

```
[root@controller ~]# . admin-openrc
[root@controller ~]# openstack user create --domain default --password-prompt glance
User Password:
Repeat User Password:
+-----------+----------------------------------+
| Field     | Value                            |
+-----------+----------------------------------+
| domain_id | d63fbba811b94cffb2cf9f88b59f4066 |
| enabled   | True                             |
| id        | 18f8b08353af43268a8caa704cc20281 |
| name      | glance                           |
+-----------+----------------------------------+
[root@controller ~]# openstack role add --project service --user glance admin
[root@controller ~]# openstack service create --name glance \
>   --description "OpenStack Image" image
+-------------+----------------------------------+
| Field       | Value                            |
+-------------+----------------------------------+
| description | OpenStack Image                  |
| enabled     | True                             |
| id          | 022a3acc4f4a486aaa79007298b3a5e9 |
| name        | glance                           |
| type        | image                            |
+-------------+----------------------------------+
```

3. 创建镜像服务的 API 端点

创建公共端点：

 # openstack endpoint create --region RegionOne image public http://controller:9292

创建外部端点：

 # openstack endpoint create --region RegionOne image internal http://controller:9292

创建管理端点：

 # openstack endpoint create --region RegionOne　image admin http://controller:9292

```
[root@controller ~]# openstack endpoint create --region RegionOne \
>    image public http://controller:9292
+--------------+----------------------------------+
| Field        | Value                            |
+--------------+----------------------------------+
| enabled      | True                             |
| id           | 525804442ff2491da09e4b4d52c34063 |
| interface    | public                           |
| region       | RegionOne                        |
| region_id    | RegionOne                        |
| service_id   | 022a3acc4f4a486aaa79007298b3a5e9 |
| service_name | glance                           |
| service_type | image                            |
| url          | http://controller:9292           |
+--------------+----------------------------------+
[root@controller ~]# openstack endpoint create --region RegionOne \
>    image internal http://controller:9292
+--------------+----------------------------------+
| Field        | Value                            |
+--------------+----------------------------------+
| enabled      | True                             |
| id           | 847e99ae0fcd402b9a1909332d476a43 |
| interface    | internal                         |
| region       | RegionOne                        |
| region_id    | RegionOne                        |
| service_id   | 022a3acc4f4a486aaa79007298b3a5e9 |
| service_name | glance                           |
| service_type | image                            |
| url          | http://controller:9292           |
+--------------+----------------------------------+
[root@controller ~]# openstack endpoint create --region RegionOne \
>    image admin http://controller:9292
+--------------+----------------------------------+
| Field        | Value                            |
+--------------+----------------------------------+
| enabled      | True                             |
| id           | 18714f11a8b340f2a02ff6cbd68144bb |
| interface    | admin                            |
| region       | RegionOne                        |
| region_id    | RegionOne                        |
| service_id   | 022a3acc4f4a486aaa79007298b3a5e9 |
| service_name | glance                           |
| service_type | image                            |
| url          | http://controller:9292           |
+--------------+----------------------------------+
```

5.4　安装并配置 Glance

1. 安装 Glance 所需软件包

 # yum install openstack-glance -y

2. 配置 Glance 所需组件

编辑**/etc/glance/glance-api.conf** 文件。

编辑**[database]**部分，配置数据库连接访问。

[database]
connection = mysql+pymysql://glance:000000@controller/glance

编辑**[keystone_authtoken]**和**[paste_deploy]**，配置 Keystone 认证服务访问。

[keystone_authtoken]
auth_uri = http://controller:5000
auth_url = http://controller:35357
memcached_servers = controller:11211
auth_type = password
project_domain_name = default

```
user_domain_name = default
project_name = service
username = glance
password = 000000

[paste_deploy]
flavor = keystone
```

注:password 密码是创建 glance 用户时设置的密码,可灵活设定。

编辑[glance_store]部分,配置本地文件系统存储和镜像位置。

```
[glance_store]
stores = file,http
default_store = file
filesystem_store_datadir = /var/lib/glance/images/
```

编辑/etc/glance/glance-registry.conf 文件。

编辑[database]部分,配置数据库连接访问。

```
[database]
connection = mysql+pymysql://glance:000000@controller/glance
```

编辑[keystone_authtoken]和[paste_deploy]部分,配置 Keystone 认证服务访问。

```
[keystone_authtoken]
auth_uri = http://controller:5000
auth_url = http://controller:35357
memcached_servers = controller:11211
auth_type = password
project_domain_name = default
user_domain_name = default
project_name = service
username = glance
password = 000000

[paste_deploy]
flavor = keystone
```

注:password 密码是创建 glance 用户时设置的密码,可灵活设定。

3. 同步数据库

```
# su -s /bin/sh -c "glance-manage db_sync" glance
```

```
[root@controller ~]# su -s /bin/sh -c "glance-manage db_sync" glance
Option "verbose" from group "DEFAULT" is deprecated for removal.  Its value may
 be silently ignored in the future.
/usr/lib/python2.7/site-packages/oslo_db/sqlalchemy/enginefacade.py:1056: OsloDB
DeprecationWarning: EngineFacade is deprecated; please use oslo_db.sqlalchemy.en
ginefacade
  expire_on_commit=expire_on_commit, _conf=conf)
/usr/lib/python2.7/site-packages/pymysql/cursors.py:146: Warning: Duplicate inde
x 'ix_image_properties_image_id_name' defined on the table 'glance.image_propert
ies'. This is deprecated and will be disallowed in a future release.
  result = self._query(query)
```

注:进入 glance 数据库查看是否有数据表,验证是否同步成功。

4. 启动 Glance 服务并设置开机自启动

systemctl enable openstack-glance-api.service openstack-glance-registry.service
systemctl start openstack-glance-api.service openstack-glance-registry.service

5.5 验证 Glance 服务

1. 生效 admin 用户环境变量

. admin-openrc

2. 下载镜像

通过网址 http://download.cirros-cloud.net/0.3.4/cirros-0.3.4-x86_64-disk.img 下载镜像，将镜像文件用 SecureFX 工具上传至控制节点"/"目录。

3. 上传镜像

使用 qcow2 磁盘格式、bare 容器格式上传镜像到镜像服务并设置公共可见，以便所有的租户都可以访问。

openstack image create "cirros" --file cirros-0.3.4-x86_64-disk.img --disk-format qcow2 --container-format bare -public

```
[root@controller ~]# openstack image create "cirros" --file cirros-0.3.4-x86_64-
disk.img --disk-format qcow2 --container-format bare --public
+------------------+------------------------------------------------------+
| Field            | value                                                |
+------------------+------------------------------------------------------+
| checksum         | 6f862a7d158576591631cb9051cbbd48                     |
| container_format | bare                                                 |
| created_at       | 2016-08-24T20:28:04Z                                 |
| disk_format      | qcow2                                                |
| file             | /v2/images/09aab83f-b55a-48c3-bb62-09ac8d2d5f8f/file |
| id               | 09aab83f-b55a-48c3-bb62-09ac8d2d5f8f                 |
| min_disk         | 0                                                    |
| min_ram          | 0                                                    |
| name             | cirros                                               |
| owner            | e71744580698483c90b1ac0a03d66313                     |
| protected        | False                                                |
| schema           | /v2/schemas/image                                    |
| size             | 13631488                                             |
| status           | active                                               |
| tags             |                                                      |
| updated_at       | 2016-08-24T20:28:05Z                                 |
| virtual_size     | None                                                 |
| visibility       | public                                               |
+------------------+------------------------------------------------------+
```

上传镜像文件的格式为：

openstack image create "cirros" --file cirros-0.3.4-x86_64-disk.img --disk-format qcow2 --container-format bare --public

其中，

"cirros"：镜像名称。

--file：上传镜像的存储位置。

--disk-format：指定镜像文件的格式，有效的为 qcow2、raw、vhd、vmdk、vdi、iso、aki、ari、ami。用户可以使用 file 命令查询文件格式。

--disk-format：指定容器的格式，有效的有 bare、ovf、aki、ari、ami。

--public：值为 true/false（可见/不可见）。

4. 查看上传的镜像并验证属性

openstack image list

```
[root@controller ~]# openstack image list
+--------------------------------------+--------+--------+
| ID                                   | Name   | Status |
+--------------------------------------+--------+--------+
| 04fa7fe2-0eb2-44b5-8f1c-8033d61a33f6 | centos | active |
+--------------------------------------+--------+--------+
```

openstack image show cirros

```
[root@controller ~]# openstack image show cirros
+------------------+------------------------------------------------------+
| Field            | Value                                                |
+------------------+------------------------------------------------------+
| checksum         | 6f862a7d158576591631cb9051cbbd48                     |
| container_format | bare                                                 |
| created_at       | 2016-08-24T20:28:04Z                                 |
| disk_format      | qcow2                                                |
| file             | /v2/images/09aab83f-b55a-48c3-bb62-09ac8d2d5f8f/file |
| id               | 09aab83f-b55a-48c3-bb62-09ac8d2d5f8f                 |
| min_disk         | 0                                                    |
| min_ram          | 0                                                    |
| name             | cirros                                               |
| owner            | e71744580698483c90b1ac0a03d66313                     |
| protected        | False                                                |
| schema           | /v2/schemas/image                                    |
| size             | 13631488                                             |
| status           | active                                               |
| tags             |                                                      |
| updated_at       | 2016-08-24T20:28:05Z                                 |
| virtual_size     | None                                                 |
| visibility       | public                                               |
+------------------+------------------------------------------------------+
```

5.6 制作 CentOS 7 镜像

第 2 章讲解了如何安装使用 KVM，使用界面方式安装了虚拟机，这里使用命令的方式来为 Glance 服务制作一个镜像。需按照第 2 章安装软件包和启用服务，镜像制作可随意在任意一台服务器中进行操作。

（1）首先确保 CentOS-7-x86_64-Minimal-1511.iso 镜像文件在服务器中，这里将该镜像文件放在/opt 目录中。

[root@storage2 ~]#**ls -l /opt/CentOS-7-x86_64-Minimal-1511.iso**

切换到/tmp 目录，创建一个 10GB 大小的镜像文件，名字为 centos7_mini。

[root@storage2 ~]#**cd /tmp/**

[root@storage2 tmp]#**qemu-img create -f raw centos7_mini.img 10G**

```
[root@storage2 ~]# ll /opt/CentOS-7-x86_64-Minimal-1511.iso
-rw-r--r--. 1 root root 632291328 Jun 10 04:30 /opt/CentOS-7-x86_64-Minimal-1511.iso
[root@storage2 ~]# cd /tmp/
[root@storage2 tmp]# qemu-img create -f raw centos7_mini.img 10G
Formatting 'centos7_mini.img', fmt=raw size=10737418240
[root@storage2 tmp]#
```

（2）启动部署虚拟机。

[root@storage2 tmp]# **virt-install --name centos7_mini --ram 1024 --vcpus=1 --disk path=/tmp/centos7_mini.img --network network:default,model=virtio --arch=x86_64 --os-type=linux --graphics vnc,port=5910 --cdrom /opt/CentOS-7-x86_64-Minimal-1511.iso --boot cdrom**

执行结果如下所示:

```
[root@storage2 tmp]# virt-install --name centos7_mini --ram 1024 --vcpus=1 --disk path=/t
mp/centos7_mini.img --network network:default,model=virtio --arch=x86_64 --os-type=linux
--graphics vnc,port=5910 --cdrom /opt/CentOS-7-x86_64-Minimal-1511.iso  --boot cdrom
WARNING  Graphics requested but DISPLAY is not set. Not running virt-viewer.
WARNING  No console to launch for the guest, defaulting to --wait -1

Starting install...
Creating domain...                                             |   0 B  00:00:00
Domain installation still in progress. Waiting for installation to complete.
```

(3) 通过 TightVNC Viewer 软件连接到虚拟桌面，安装虚拟机，如图 5-1 所示。

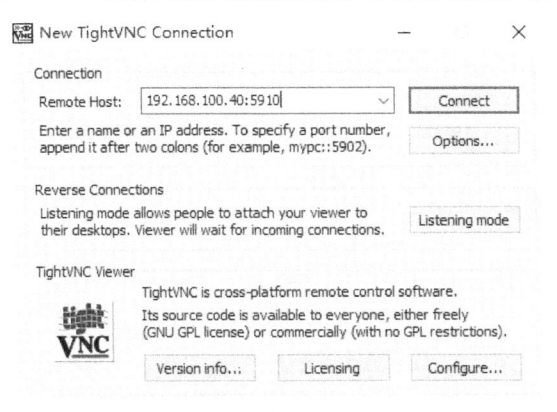

图 5-1 连接虚拟主机

在如图 5-1 所示的对话框中输入 IP，将端口号连接到虚拟桌面。

(4) 开始安装虚拟操作系统。

第一步：安装 CentOS 7 操作系统（如图 5-2 所示）。

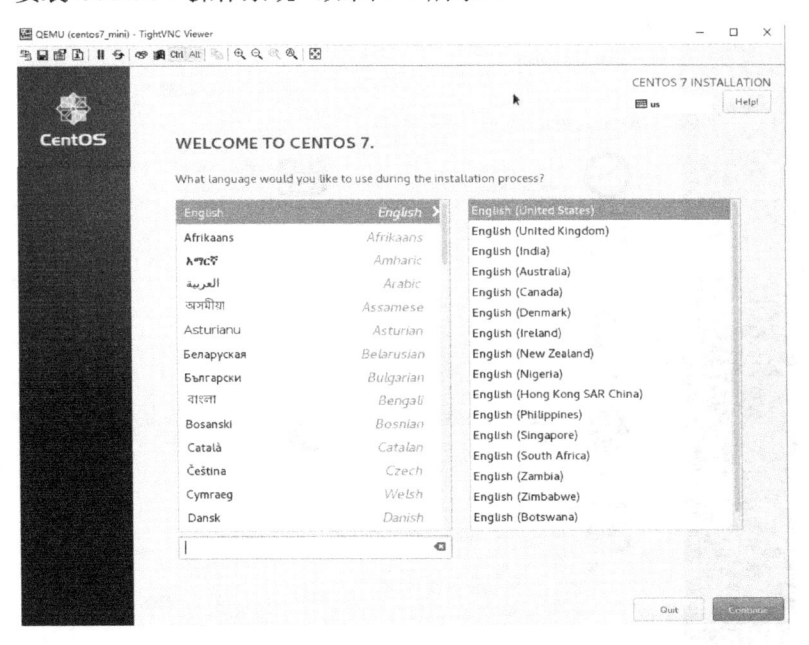

图 5-2 安装 CentOS 7 操作系统

第二步：更改磁盘（如图 5-3 所示）。

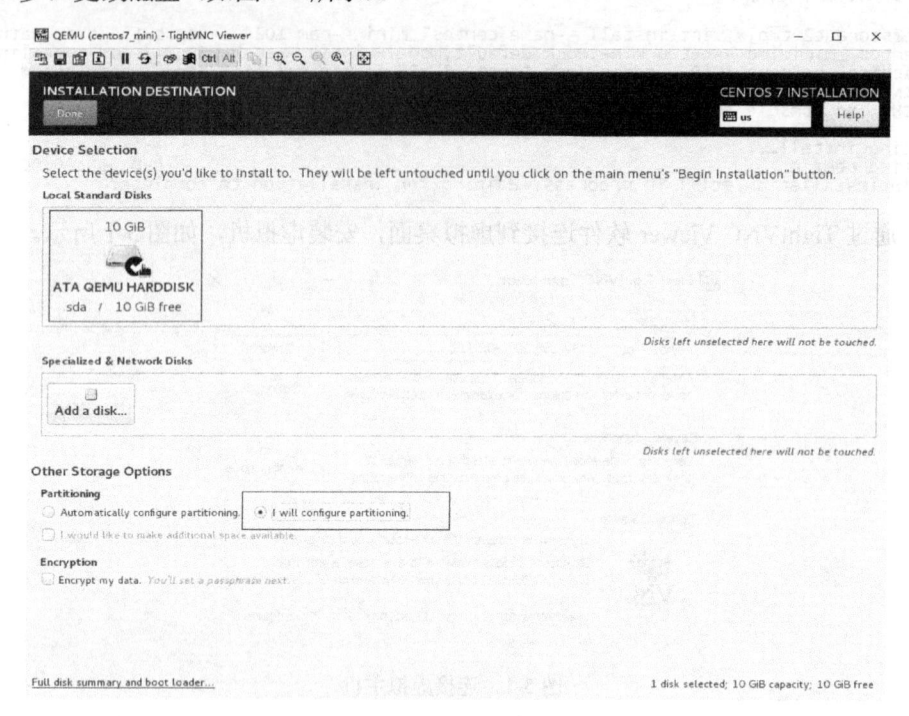

图 5-3　更改磁盘

第三步：选择磁盘（如图 5-4 所示）。

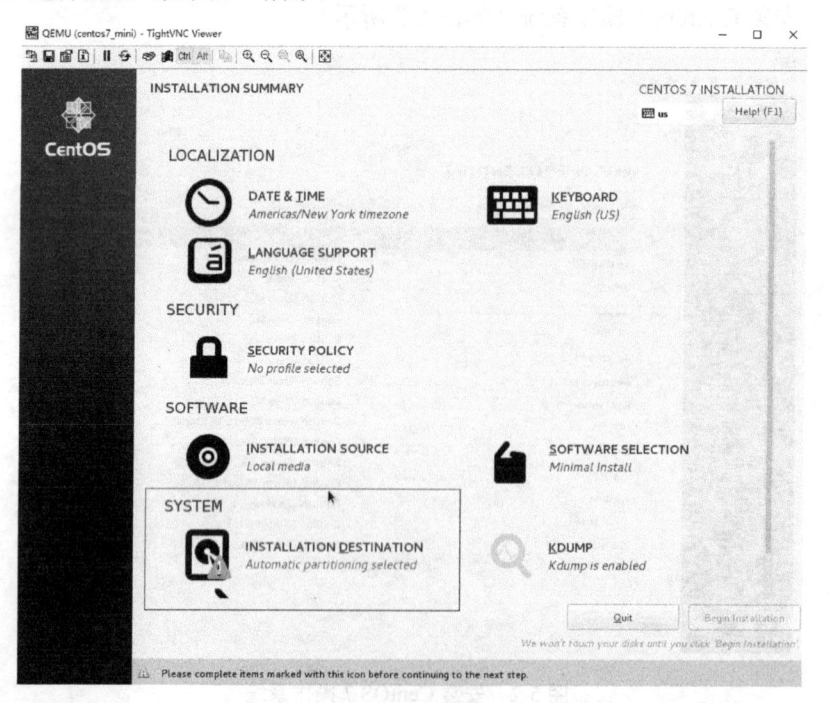

图 5-4　选择磁盘

第四步：磁盘分区操作（如图 5-5 所示）。

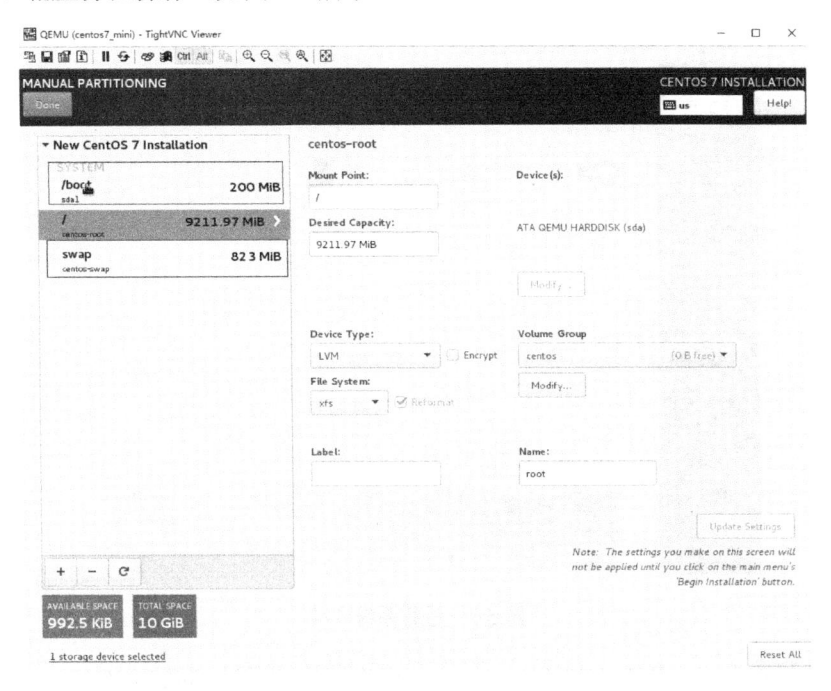

图 5-5　磁盘分区操作

第五步：设置 Root 账户密码（如图 5-6 所示）。

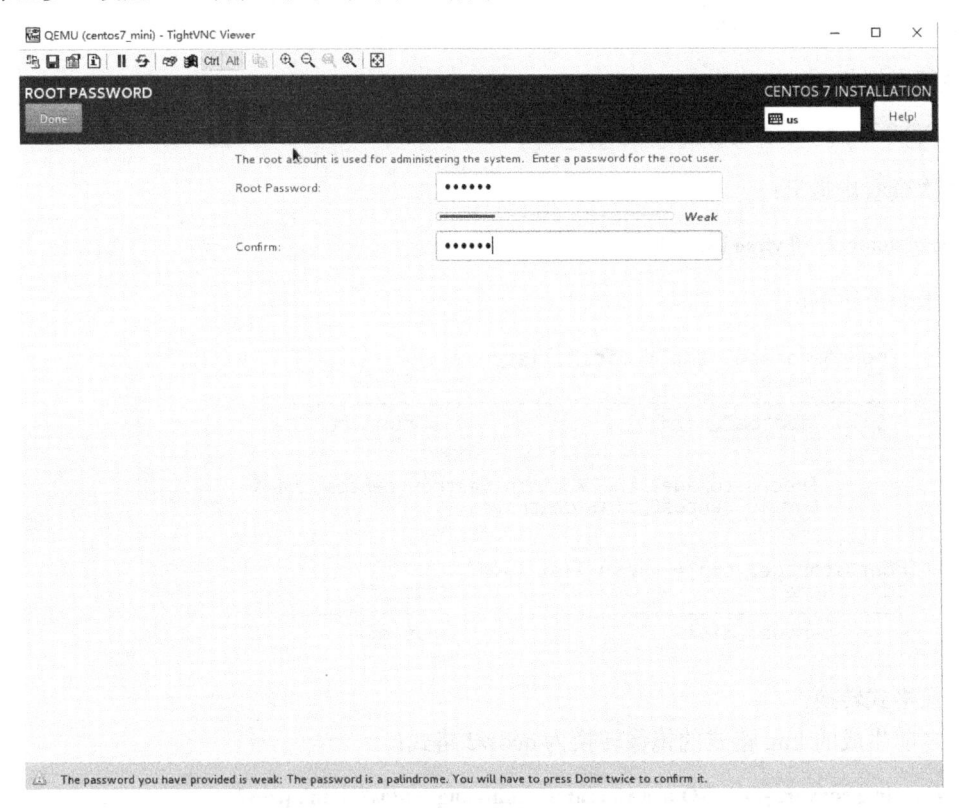

图 5-6　设置 Root 账户密码

第六步：进行系统安装（如图 5-7 所示）。

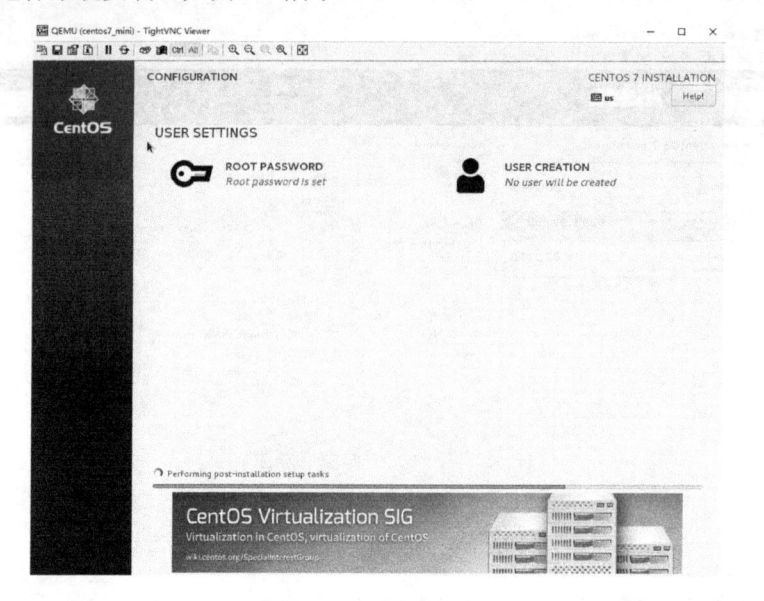

图 5-7 进行系统安装

完成安装重启之后，查看虚拟机状态，启动虚拟机重新通过 TightVNC Viewer 软件连接到虚拟桌面。

查看虚拟机状态：

[root@storage2 ~]# **virsh list --all**

关闭虚拟机：

[root@storage2 ~]# **virsh destroy centos7_mini**

再次查看虚拟机：

[root@storage2 ~]# **virsh list --all**

结果如下所示：

```
[root@storage2 tmp]# virsh list --all
 Id    Name                           State
----------------------------------------------------
 3     centos7_mini                   running

[root@storage2 tmp]# virsh destroy centos7_mini
Domain centos7_mini destroyed

[root@storage2 tmp]# virsh list --all
 Id    Name                           State
----------------------------------------------------
 -     centos7_mini                   shut off
```

镜像格式转换。

将之前生成的 img 格式的镜像转换为 qcow2 格式：

qemu-img convert -f raw -O qcow2 centos7_mini.img centos7_mini.qcow2

转换完成之后，将会出现之前在 Glance 组件中使用的镜像，这样就可以像 5.5 节中一样，上传一个属于自己的镜像。

课后习题

1. _____ 是 OpenStack 镜像服务，用来注册、登录和检索虚拟机镜像。
2. 创建一个名为 glance 的用户的命令是_____。
3. Glance 管理端点的端口号是_____。
4. 同步 Glance 数据库的命令是_____。
5. 上传镜像 cirros-0.3.4-x86_64-disk.img 并将镜像名命名为 CIRROS，如何实现？

第6章 计算服务 Nova

6.1 Nova 架构及原理

Nova 是 OpenStack 云中的计算组织控制器。OpenStack 中云主机（instances）生命周期的所有活动都由 Nova 处理。这样使得 Nova 成为一个负责管理计算资源、网络、认证、所需可扩展性的平台。但是，Nova 自身并没有提供任何虚拟化能力，而是使用 libvirt API 来与被支持的 Hypervisors 交互。OpenStack 计算服务（Nova）由下列组件构成。

（1）API Server

对外提供一个与云基础设施交互的接口，也是外部可用于管理基础设施的唯一组件。管理使用 EC2 API 通过 Web Services 调用实现。然后，API Server 通过消息队列（Message Queue）轮流与云基础设施的相关组件通信。作为 EC2 API 的另外一种选择，OpenStack 也提供一个内部使用的"OpenStack API"。

（2）Message Queue（RabbitMQ Server）

OpenStack 节点之间通过消息队列使用 AMQP（Advanced Message Queue Protocol，高级消息队列协议）完成通信。Nova 通过异步调用请求响应，使用回调函数在收到响应时触发。因为使用了异步通信，所以不会有用户长时间卡在等待状态。这是有效的，因为许多 API 调用预期的行为都非常耗时，例如加载一个云主机，或者上传一个镜像。

（3）Compute Worker（Nova-compute）

Compute Worker 处理管理云主机生命周期。它们通过消息服务接收云主机生命周期管理的请求，并承担操作工作。在一个典型生产环境的云部署中有一些 Compute Workers。云主机部署在哪个可用的 Compute Worker 上取决于调度算法。

（4）Network Controller（Nova-network）

Network Controller 处理主机地网络配置。它包括 IP 地址分配、为项目配置 VLAN、实现安全组、配置计算节点网络。

(5) Volume Workers（Nova-volume）

Volume Workers 用来管理基于 LVM（Logical Volume Manager，逻辑卷管理）的云主机卷。Volume Workers 有卷的相关功能，例如新建卷，删除卷，为云主机附加卷，为云主机分离卷。卷为云主机提供一个持久化存储，因为根分区是非持久化的，所以当云主机终止时对它所做的任何改变都会丢失。当一个卷从云主机分离或者云主机终止（这个卷附加在该终止的云主机上）时，这个卷保留着存储在其上的数据。当把这个卷重新附加到相同云主机或者附加到不同云主机上时，这些数据依旧能被访问。

一个云主机的重要数据几乎总是要写在卷上，这样可以确保能在以后访问。这个对存储的典型应用需要数据库等服务的支持。

(6) Scheduler（Nova-scheduler）

调度器 Scheduler 把 Nova-API 调用映射为 OpenStack 组件。调度器作为一个称为 Nova-schedule 的守护进程运行，通过恰当的调度算法从可用资源池获得一个计算服务。Scheduler 会根据负载、内存、可用域的物理距离、CPU 构架等做出调度决定。Nova-schedule 实现了一个可插入式的结构。

当前，Nova-scheduler 实现了一些基本的调度算法：①随机算法，计算主机在所有可用域内随机选择；②可用域算法，与随机算法相仿，但是计算主机在指定的可用域内随机选择；③简单算法，这种方法选择负载最小的主机运行云主机。负载信息可通过负载均衡器获得。

总的来说，Nova 的各个组件（如图 6-1 所示）是以数据库和队列为中心进行通信的，下面对其中的几个组件做简单的介绍。

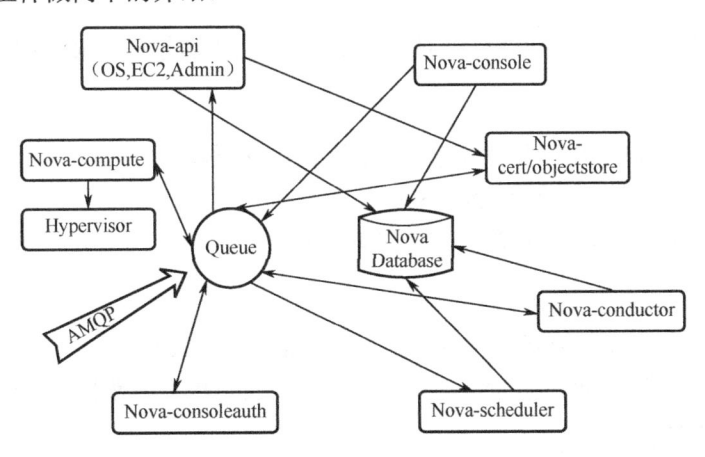

图 6-1　Nova 的各个组件

(1) API

nova-api

接收和响应客户的 API 调用。

(2) Console Interface

nova-console

用户可以通过多种方式访问虚拟机的控制台：

Nova-novncproxy，基于 Web 浏览器的 VNC 访问 Nova-spicehtml5proxy，基于 HTML5 浏

览器的 SPICE 访问 Nova-xvpnvncproxy，基于 Java 客户端的 VNC 访问：

nova-consoleauth

负责对访问虚拟机控制台提供 Token 认证：

nova-cert

提供 x509 证书支持。

（3）Database

Nova 会有一些数据需要存放到数据库中，一般使用 MySQL。

一般数据库安装在控制节点上。

（4）Message Queue

前面介绍了 Nova 包含众多的子服务，这些子服务之间需要相互协调和通信。

为解耦各个子服务，Nova 通过 Message Queue 作为子服务的信息中转站。

所以在架构图上我们看到了子服务之间没有直接的连线，它们都通过 Message Queue 联系。

OpenStack 默认是用 RabbitMQ 作为 Message Queue。

（5）Compute Core

nova-scheduler

虚拟机调度服务，负责决定在哪个计算节点上运行虚拟机。

nova-compute

管理虚拟机的核心服务，通过调用 Hypervisor API 实现虚拟机生命周期管理。

Hypervisor

计算节点上跑的虚拟化管理程序，虚拟机管理底层的程序。

不同虚拟化技术提供自己的 Hypervisor。

常用的 Hypervisor 有 KVM、Xen、VMWare 等。

nova-conductor

Nova-compute 经常需要更新数据库，比如更新虚拟机的状态。

出于安全性和伸缩性的考虑，Nova-compute 并不会直接访问数据库，而是将这个任务委托给 Nova-conductor。

6.2 安装并配置控制节点

6.2.1 数据库配置

登录 MySQL 并创建 nova_api 和 nova 数据库：

mysql -uroot -p000000

创建 nova_api、nova 数据库：

MariaDB [(none)]>**CREATE DATABASE nova_api;**
MariaDB [(none)]>**CREATE DATABASE nova;**

设置授权用户和密码：

MariaDB [(none)]>**GRANT ALL PRIVILEGES ON nova_api.* TO 'nova'@'localhost' IDENTIFIED BY '000000';**

MariaDB [(none)]> **GRANT ALL PRIVILEGES ON nova_api.* TO 'nova'@'%' IDENTIFIED BY '000000';**

MariaDB [(none)]> **GRANT ALL PRIVILEGES ON nova.* TO 'nova'@'localhost' IDENTIFIED BY '000000';**

MariaDB [(none)]>**GRANT ALL PRIVILEGES ON nova.* TO 'nova'@'%' IDENTIFIED BY '000000';**

6.2.2 创建服务凭证和 API 端点

1. 生效 admin 用户环境变量

 # . admin-openrc

2. 创建服务凭证

创建名为 nova 的用户（user）：

 # openstack user create --domain default --password-prompt nova

进行关联：给 nova 用户添加 admin 角色：

 # openstack role add --project service --user nova admin

创建 nova 服务实体认证：

 # openstack service create --name nova --description "OpenStack Compute" compute

```
[root@controller ~]# . admin-openrc
[root@controller ~]# openstack user create --domain default \
>   --password-prompt nova
User Password:
Repeat User Password:
+-----------+----------------------------------+
| Field     | Value                            |
+-----------+----------------------------------+
| domain_id | d63fbba811b94cffb2cf9f88b59f4066 |
| enabled   | True                             |
| id        | 67cb99a8c47c44cf9e6cbcf97d5fb117 |
| name      | nova                             |
+-----------+----------------------------------+
[root@controller ~]# openstack role add --project service --user nova admin
[root@controller ~]# openstack service create --name nova \
>   --description "OpenStack Compute" compute
+-------------+----------------------------------+
| Field       | Value                            |
+-------------+----------------------------------+
| description | OpenStack Compute                |
| enabled     | True                             |
| id          | 22aef80683e54200abe9bdd75016be40 |
| name        | nova                             |
| type        | compute                          |
+-------------+----------------------------------+
```

3. 创建 API 端点

创建公共端点：

 # openstack endpoint create --region RegionOne compute public http://controller:8774/v2.1/%\(tenant_id\)s

创建外部端点：

openstack endpoint create --region RegionOne compute internal http://controller:8774/v2.1/%\(tenant_id\)s

创建管理端点：

openstack endpoint create --region RegionOne compute admin http://controller:8774/v2.1/%\(tenant_id\)s

```
[root@controller ~]# openstack endpoint create --region RegionOne \
>   image public http://controller:9292
+--------------+----------------------------------+
| Field        | Value                            |
+--------------+----------------------------------+
| enabled      | True                             |
| id           | 525804442ff2491da09e4b4d52c34063 |
| interface    | public                           |
| region       | RegionOne                        |
| region_id    | RegionOne                        |
| service_id   | 022a3acc4f4a486aaa79007298b3a5e9 |
| service_name | glance                           |
| service_type | image                            |
| url          | http://controller:9292           |
+--------------+----------------------------------+
[root@controller ~]# openstack endpoint create --region RegionOne \
>   image internal http://controller:9292
+--------------+----------------------------------+
| Field        | Value                            |
+--------------+----------------------------------+
| enabled      | True                             |
| id           | 847e99ae0fcd402b9a1909332d476a43 |
| interface    | internal                         |
| region       | RegionOne                        |
| region_id    | RegionOne                        |
| service_id   | 022a3acc4f4a486aaa79007298b3a5e9 |
| service_name | glance                           |
| service_type | image                            |
| url          | http://controller:9292           |
+--------------+----------------------------------+
[root@controller ~]# openstack endpoint create --region RegionOne \
>   image admin http://controller:9292
+--------------+----------------------------------+
| Field        | Value                            |
+--------------+----------------------------------+
| enabled      | True                             |
| id           | 18714f11a8b340f2a02ff6cbd68144bb |
| interface    | admin                            |
| region       | RegionOne                        |
| region_id    | RegionOne                        |
| service_id   | 022a3acc4f4a486aaa79007298b3a5e9 |
| service_name | glance                           |
| service_type | image                            |
| url          | http://controller:9292           |
+--------------+----------------------------------+
```

6.2.3 安装并配置 Nova 组件

1. 安装 Nova 组件所需软件包

yum install openstack-nova-api openstack-nova-conductor openstack-nova-console openstack-nova-novncproxy openstack-nova-scheduler -y

2. 配置 Nova 所需组件

编辑 **/etc/nova/nova.conf** 文件。

编辑 **[DEFAULT]** 部分，启用计算和元数据 API。

[DEFAULT]
enabled_apis = osapi_compute,metadata

编辑[api_database]和[database]部分，配置数据库链接。

[api_database]
connection = mysql+pymysql://nova:000000@controller/nova_api
[database]
connection = mysql+pymysql://nova:000000@controller/nova

编辑[DEFAULT]和[oslo_messaging_rabbit]部分，配置 RabbitMQ 消息服务器链接。

[DEFAULT]
rpc_backend = rabbit

[oslo_messaging_rabbit]
rabbit_host = controller
rabbit_userid = openstack
rabbit_password = 000000

编辑[DEFAULT]和[keystone_authtoken]部分，配置 Keystone 身份认证。

[DEFAULT]
auth_strategy = keystone

[keystone_authtoken]
auth_uri = http://controller:5000
auth_url = http://controller:35357
memcached_servers = controller:11211
auth_type = password
project_domain_name = default
user_domain_name = default
project_name = service
username = nova
password = 000000

编辑[DEFAULT]部分，配置管理 IP 地址和启用网络服务。

[DEFAULT]
my_ip = 192.168.100.10
use_neutron = True
firewall_driver = nova.virt.firewall.NoopFirewallDriver

编辑[vnc]部分，配置 VNC 代理管理 IP 地址。

[vnc]
vncserver_listen = $my_ip
vncserver_proxyclient_address = $my_ip

编辑[glance]部分，配置镜像服务 API 端点。

[glance]
api_servers = http://controller:9292

编辑[oslo_concurrency]部分，配置 loca_path。

```
[oslo_concurrency]
lock_path = /var/lib/nova/tmp
```

3. 同步数据库

```
# su -s /bin/sh -c "nova-manage api_db sync" nova
# su -s /bin/sh -c "nova-manage db sync" nova
```

```
[root@controller ~]# su -s /bin/sh -c "nova-manage api_db sync" nova
[root@controller ~]# su -s /bin/sh -c "nova-manage db sync" nova
/usr/lib/python2.7/site-packages/pymysql/cursors.py:146: Warning: Duplicate inde
x 'block_device_mapping_instance_uuid_virtual_name_device_name_idx' defined on t
he table 'nova.block_device_mapping'. This is deprecated and will be disallowed
in a future release.
  result = self._query(query)
/usr/lib/python2.7/site-packages/pymysql/cursors.py:146: Warning: Duplicate inde
x 'uniq_instances0uuid' defined on the table 'nova.instances'. This is deprecate
d and will be disallowed in a future release.
  result = self._query(query)
```

注：进入 nova 数据库查看是否有数据表，验证是否同步成功。

4. 启动 Nova 服务并设置开机自启动

```
# systemctl enable openstack-nova-api.service openstack-nova-consoleauth.service openstack-nova-scheduler.service openstack-nova-conductor.service openstack-nova-novncproxy.service
# systemctl start openstack-nova-api.service openstack-nova-consoleauth.service openstack-nova-scheduler.service openstack-nova-conductor.service openstack-nova-novncproxy.service
```

6.3 安装并配置计算节点

6.3.1 安装并配置 Nova 组件

1. 安装 Nova 组件所需软件包

```
# yum install openstack-nova-compute -y
```

2. 配置 Nova 所需组件

编辑 **/etc/nova/nova.conf** 文件。

编辑 **[DEFAULT]** 和 **[oslo_messaging_rabbit]** 部分，配置 RabbitMQ 消息服务器链接。

```
[DEFAULT]
rpc_backend = rabbit

[oslo_messaging_rabbit]
rabbit_host = controller
rabbit_userid = openstack
rabbit_password = 000000
```

编辑 **[DEFAULT]** 和 **[keystone_authtoken]** 部分，配置 Keystone 身份认证。

```
[DEFAULT]
auth_strategy = keystone


[keystone_authtoken]
```

```
auth_uri = http://controller:5000
auth_url = http://controller:35357
memcached_servers = controller:11211
auth_type = password
project_domain_name = default
user_domain_name = default
project_name = service
username = nova
password = 000000
```

编辑[DEFAULT]部分,配置管理 IP 地址和启用网络服务支持。

```
[DEFAULT]
my_ip = 192.168.100.20
use_neutron = True
firewall_driver = nova.virt.firewall.NoopFirewallDriver
```

编辑[vnc]部分,启用并配置远程控制台的访问。

```
[vnc]
enabled = True
vncserver_listen = 0.0.0.0
vncserver_proxyclient_address = $my_ip
novncproxy_base_url = http://controller:6080/vnc_auto.html
```

编辑[glance]部分,配置镜像服务 API 的位置。

```
[glance]
api_servers = http://controller:9292
```

编辑[oslo_concurrency]部分,配置 loca_path。

```
[oslo_concurrency]
lock_path = /var/lib/nova/tmp
```

6.3.2 检查主机是否支持虚拟机硬件加速

1. 执行命令

```
# egrep -c '(vmx|svm)' /proc/cpuinfo
```

注:
如果该命令返回一个 1 或更大的值,说明你的系统支持硬件加速,通常不需要额外的配置。
如果这个指令返回一个 0 值,说明你的系统不支持硬件加速,你必须配置 libvirt 取代 KVM 来使用 QEMU。
如果是虚拟机返回的是 0,则不需要修改。
编辑**/etc/nova/nova.conf** 文件。
编辑[libvirt]部分:

```
[libvirt]
virt_type = qemu
```

2. 启动 Nova 服务并设置开机自启动

```
# systemctl enable libvirtd.service openstack-nova-compute.service
# systemctl start libvirtd.service openstack-nova-compute.service
```

6.4 验证 Nova 服务

在控制节点进行如下操作。

1. 生效 admin 用户环境变量

```
# . admin-openrc
```

2. 查看 Nova 服务

```
# openstack compute service list
```

```
[root@controller ~]# . admin-openrc
[root@controller ~]# openstack compute service list
+----+------------------+------------+----------+---------+-------+----------------------------+
| Id | Binary           | Host       | Zone     | Status  | State | Updated At                 |
+----+------------------+------------+----------+---------+-------+----------------------------+
| 1  | nova-            | controller | internal | enabled | up    | 2016-08-24T1               |
|    | consoleauth      |            |          |         |       | 8:59:01.0000               |
|    |                  |            |          |         |       | 00                         |
| 2  | nova-            | controller | internal | enabled | up    | 2016-08-24T1               |
|    | conductor        |            |          |         |       | 8:59:01.0000               |
|    |                  |            |          |         |       | 00                         |
| 3  | nova-            | controller | internal | enabled | up    | 2016-08-24T1               |
|    | scheduler        |            |          |         |       | 8:59:01.0000               |
|    |                  |            |          |         |       | 00                         |
| 6  | nova-compute     | compute    | nova     | enabled | up    | 2016-08-24T1               |
|    |                  |            |          |         |       | 8:59:01.0000               |
|    |                  |            |          |         |       | 00                         |
+----+------------------+------------+----------+---------+-------+----------------------------+
```

课后习题

1. OpenStack 计算服务（Nova）由以下组件所构成：_____、_____、_____、_____、_____、_____。
2. Nova 的各个组件是以_____和_____为中心进行通信的。
3. 同步 Nova-api 数据库的命令是_____。
4. 检查主机是否支持虚拟机硬件加速的命令是_____。
5. 启动 Nova 计算节点 Nova 相关服务的命令是_____。

第7章 网络部署服务 Neutron

7.1 Neutron 基本概念

（1）Openstack 网络（Neutron）概念：OpenStack 网络（Neutron）管理 OpenStack 环境中虚拟网络基础设施（VNI）的所有方面和物理网络基础设施（PNI）的接入层方面。OpenStack Networking 允许租户创建高级虚拟网络拓扑，包括防火墙，负载均衡和虚拟私有网络（VPN）等服务。

（2）网络服务提供网络、子网和路由对象的概念。每个概念有自己的功能，可以模仿对应的物理设备：网络包括子网，路由则在不同的子网和网络之间进行路由转发。每个路由都有一个连接到网络的网关，并且很多接口都连接到子网中。子网可以访问其他连接到相同路由的其他子网的机器。

（3）任何给定的 Networking 设置至少有一个外部网络（简称外网）。不像其他的网络，外部网络不仅仅是一个虚拟定义的网络。相反，它代表了一种 OpenStack 安装之外的能从物理的，外部的网络访问的视图。外部网络上的 IP 地址能被任何物理接入外面网络的人所访问。因为外部网络仅仅代表了进入外面网络的一个视图，网络上的 DHCP 是关闭的。

（4）外部网络之外，任何 Networking 设置拥有一个或多个内部网络。这些软件定义的网络直接连接到虚拟机。仅仅在给定网络上的虚拟机，或那些在通过接口连接到相近路由的子网上的虚拟机，能直接访问连接到那个网络上的虚拟机。如果外网需要访问虚拟机，或者相反，则网络中的路由器就是必须使用的。每个路由器配有一个网关，可以连接到网络和接口，这些接口又连接着子网。如同实体路由器一样，子网中的机器可以访问连接到同一个路由器的子网中的其他机器，机器可以通过该路由器的网关访问外网。

（5）能够将外部网络的 IP 地址分配到内部网络的端口。无论何时一旦有连接连接到子网，这个连接就被称为一个端口。能连接外部网络的 IP 地址和虚拟机的端口。这样，外部网络的实体就能访问虚拟机了。

（6）Networking 也支持安全组。安全组允许管理员分组定义防火墙规则。一个虚拟机可以属于一个或多个安全组，Networking 针对这个虚拟机，应用这些安全组里的规则来阻塞或者开启端口、端口范围、或通信类型。

（7）每一个 Networking 使用的插件都有其自有的概念。虽然对操作 VNI 和 OpenStack 环境不是至关重要的，但理解这些概念能帮助你设置 Networking。所有的 Networking 安装使用了一个核心插件和一个安全组插件（或仅是空操作安全组插件）。另外，防火墙即服务（FWaaS）和负载均衡即服务（LBaaS）插件是可用的。

7.2 安装并配置控制节点

7.2.1 数据库配置

登录 MySQL 并创建 neutron 数据库：

`# mysql -uroot -p000000`

创建 neutron 数据库：

`MariaDB [(none)]>CREATE DATABASE neutron;`

设置授权用户和密码：

`MariaDB [(none)]>GRANT ALL PRIVILEGES ON neutron.* TO 'neutron'@'localhost' IDENTIFIED BY '000000';`

`MariaDB [(none)]>GRANT ALL PRIVILEGES ON neutron.* TO 'neutron'@'%' IDENTIFIED BY '000000';`

`MariaDB [(none)]>exit`

7.2.2 创建服务凭证和 API 端点

1. 生效 admin 用户环境变量

`# . admin-openrc`

2. 创建服务凭证

创建名为 neutron 的 user：

`# openstack user create --domain default --password-prompt neutron`

进行关联

`# openstack role add --project service --user neutron admin`

创建 Neutron 服务实体认证：

`# openstack service create --name neutron --description "OpenStack Networking" network`

```
[root@controller ~]# openstack user create --domain default --password-prompt neu
tron
User Password:
Repeat User Password:
+-----------+----------------------------------+
| Field     | Value                            |
+-----------+----------------------------------+
| domain_id | d63fbba811b94cffb2cf9f88b59f4066 |
| enabled   | True                             |
| id        | 509d25faaaeb4e99a944a9327f1a5957 |
| name      | neutron                          |
+-----------+----------------------------------+
[root@controller ~]# openstack role add --project service --user neutron admin
[root@controller ~]# openstack service create --name neutron \
>   --description "OpenStack Networking" network
+-------------+----------------------------------+
| Field       | Value                            |
+-------------+----------------------------------+
| description | OpenStack Networking             |
| enabled     | True                             |
| id          | 23ec34418ea3405da2a040e3d92576b7 |
| name        | neutron                          |
| type        | network                          |
+-------------+----------------------------------+
```

3. 创建 API 端点

创建公共端点：

 # openstack endpoint create --region RegionOne network public http://controller:9696

创建外部端点：

 # openstack endpoint create --region RegionOne network internal http://controller:9696

创建管理端点：

 # openstack endpoint create --region RegionOne network admin http://controller:9696

```
[root@controller ~]# openstack endpoint create --region RegionOne \
>   network public http://controller:9696
+--------------+----------------------------------+
| Field        | Value                            |
+--------------+----------------------------------+
| enabled      | True                             |
| id           | fdf31c0409264970880e25f90a9d5786 |
| interface    | public                           |
| region       | RegionOne                        |
| region_id    | RegionOne                        |
| service_id   | 23ec34418ea3405da2a040e3d92576b7 |
| service_name | neutron                          |
| service_type | network                          |
| url          | http://controller:9696           |
+--------------+----------------------------------+
[root@controller ~]# openstack endpoint create --region RegionOne \
>   network internal http://controller:9696
+--------------+----------------------------------+
| Field        | Value                            |
+--------------+----------------------------------+
| enabled      | True                             |
| id           | 68ab39ce72ac422ebac93adf9c942624 |
| interface    | internal                         |
| region       | RegionOne                        |
| region_id    | RegionOne                        |
| service_id   | 23ec34418ea3405da2a040e3d92576b7 |
| service_name | neutron                          |
| service_type | network                          |
| url          | http://controller:9696           |
+--------------+----------------------------------+
[root@controller ~]# openstack endpoint create --region RegionOne \
>   network admin http://controller:9696
+--------------+----------------------------------+
| Field        | Value                            |
+--------------+----------------------------------+
| enabled      | True                             |
| id           | 7d6dbb79fa4741fbb9398a063878cefe |
| interface    | admin                            |
| region       | RegionOne                        |
| region_id    | RegionOne                        |
| service_id   | 23ec34418ea3405da2a040e3d92576b7 |
| service_name | neutron                          |
| service_type | network                          |
| url          | http://controller:9696           |
+--------------+----------------------------------+
```

7.2.3 安装并配置 Neutron 组件

1. 安装 Neutron 组件所需软件包

`# yum install openstack-neutron openstack-neutron-ml2 openstack-neutron-linuxbridge ebtables -y`

2. 配置 Neutron 所需组件

编辑 **/etc/neutron/neutron.conf** 文件。

编辑 **[database]** 部分，配置数据库链接。

```
[database]
connection = mysql+pymysql://neutron:000000@controller/neutron
```

编辑 **[DEFAULT]** 部分，配置模块化 ML2 插件。

```
[DEFAULT]
core_plugin = ml2
service_plugins = router
allow_overlapping_ips = True
```

编辑 **[DEFAULT]** 和 **[oslo_messaging_rabbit]** 部分，配置 RabbitMQ 消息服务器链接。

```
[DEFAULT]
rpc_backend = rabbit

[oslo_messaging_rabbit]
rabbit_host = controller
rabbit_userid = openstack
rabbit_password = 000000
```

编辑 **[DEFAULT]** 和 **[keystone_authtoken]** 部分，配置 Keystone 身份认证。

```
[DEFAULT]
auth_strategy = keystone

[keystone_authtoken]
auth_uri = http://controller:5000
auth_url = http://controller:35357
memcached_servers = controller:11211
auth_type = password
project_domain_name = default
user_domain_name = default
project_name = service
username = neutron
password = 000000
```

编辑 **[DEFAULT]** 和 **[nova]** 部分，配置网络来通知网络拓扑结构的变化。

```
[DEFAULT]
notify_nova_on_port_status_changes = True
notify_nova_on_port_data_changes = True
```

```
[nova]
auth_url = http://controller:35357
auth_type = password
project_domain_name = default
user_domain_name = default
region_name = RegionOne
project_name = service
username = nova
password = 000000
```

编辑[oslo_concurrency]部分,配置 loca_path。

```
[oslo_concurrency]
lock_path = /var/lib/neutron/tmp
```

3. 配置 ML2 插件

编辑/etc/neutron/plugins/ml2/ml2_conf.ini 文件。

编辑[ml2]部分,配置 Flat、VLAN、VxLAN 网络。

```
[ml2]
type_drivers = flat,vlan,vxlan
```

编辑[ml2]部分,使用 VxLAN 网络。

```
[ml2]
tenant_network_types = vxlan
```

编辑[ml2]部分,启用网桥和 ML2 入口机制。

```
[ml2]
mechanism_drivers = linuxbridge,l2population
```

编辑[ml2]部分,启用端口安全扩展驱动程序。

```
[ml2]
extension_drivers = port_security
```

编辑[ml2_type_flat]部分,配置虚拟网络为 Flat 网络。

```
[ml2_type_flat]
flat_networks = provider
```

编辑[ml2_type_vxlan]部分,配置 VxLAN 网络标识符范围。

```
[ml2_type_vxlan]
vni_ranges = 1:1000
```

编辑[securitygroup]部分,配置 ipset 安全组规则。

```
[securitygroup]
enable_ipset = True
```

4. 配置 linux bridge 插件

编辑/etc/neutron/plugins/ml2/linuxbridge_agent.ini 文件。

编辑**[linux_bridge]**部分，配置虚拟网络映射到物理网络接口。

[linux_bridge]
physical_interface_mappings = provider:enp1s0f1 （物理机的外网网卡名）

编辑**[vxlan]**部分，使 VxLAN 覆盖网络，并配置物理网络的 IP 地址。

[vxlan]
enable_vxlan = True
local_ip = 192.168.100.10
l2_population = True

编辑**[securitygroup]**部分，配置安全组和网桥配置防火墙驱动。

[securitygroup]
enable_security_group = True
firewall_driver = neutron.agent.linux.iptables_firewall.IptablesFirewallDriver

5. 配置 L3 插件

编辑**/etc/neutron/l3_agent.ini** 文件。

编辑**[DEFAULT]**部分，配置网桥接口驱动和外部网络连接。

[DEFAULT]
interface_driver = neutron.agent.linux.interface.BridgeInterfaceDriver
external_network_bridge =

注：external_network_bridge（这里缺少一个值，为空值）。

6. 配置 DHCP 插件

编辑**/etc/neutron/dhcp_agent.ini** 文件。

编辑**[DEFAULT]**部分，配置网桥接口驱动、Dnsmasq DHCP 的驱动，并启用 Metadata。

[DEFAULT]
interface_driver = neutron.agent.linux.interface.BridgeInterfaceDriver
dhcp_driver = neutron.agent.linux.dhcp.Dnsmasq
enable_isolated_metadata = True

7. 配置 metadata 插件

编辑**/etc/neutron/metadata_agent.ini** 文件。

编辑**[DEFAULT]**部分，配置元数据主机和共享密钥。

nova_metadata_ip = controller
metadata_proxy_shared_secret = 000000   #metadata 代理密钥，自定义

8. 配置 Nova 服务使用网络

编辑**/etc/nova/nova.conf** 文件。

编辑**[neutron]**部分，配置访问参数，并启用和配置代理。

[neutron]
url = http://controller:9696
auth_url = http://controller:35357
auth_type = password
project_domain_name = default

```
user_domain_name = default
region_name = RegionOne
project_name = service
username = neutron
password = 000000          #创建 neutron 用户的密码
service_metadata_proxy = True
metadata_proxy_shared_secret = 000000   #metadata 代理密钥
```

9. 创建软链接

ln -s /etc/neutron/plugins/ml2/ml2_conf.ini /etc/neutron/plugin.ini

10. 同步数据库

su -s /bin/sh -c "neutron-db-manage --config-file /etc/neutron/neutron.conf --config-file /etc/neutron/plugins/ml2/ml2_conf.ini upgrade head" neutron

```
[root@controller ~]# su -s /bin/sh -c "neutron-db-manage --config-file /etc/neutron/neutron.conf \
--config-file /etc/neutron/plugins/ml2/ml2_conf.ini upgrade head" neutron
INFO  [alembic.runtime.migration] Context impl MySQLImpl.
INFO  [alembic.runtime.migration] Will assume non-transactional DDL.
  Running upgrade for neutron ...
INFO  [alembic.runtime.migration] Context impl MySQLImpl.
INFO  [alembic.runtime.migration] Will assume non-transactional DDL.
INFO  [alembic.runtime.migration] Running upgrade  -> kilo, kilo_initial
INFO  [alembic.runtime.migration] Running upgrade kilo -> 354db87e3225, nsxv_vdr_m
etadata.py
INFO  [alembic.runtime.migration] Running upgrade 354db87e3225 -> 599c6a226151, ne
utrondb_ipam
INFO  [alembic.runtime.migration] Running upgrade 599c6a226151 -> 52c5312f6baf, In
itial operations in support of address scopes
INFO  [alembic.runtime.migration] Running upgrade 52c5312f6baf -> 313373c0ffee, Fl
avor framework
INFO  [alembic.runtime.migration] Running upgrade 313373c0ffee -> 8675309a5c4f, network_rbac
INFO  [alembic.runtime.migration] Running upgrade 8675309a5c4f -> 45f955889773, quota_usage
INFO  [alembic.runtime.migration] Running upgrade 45f955889773 -> 26c371498592, subnetpool hash
INFO  [alembic.runtime.migration] Running upgrade 26c371498592 -> 1c844d1677f7, add order to dnsnameservers
INFO  [alembic.runtime.migration] Running upgrade 1c844d1677f7 -> 1b4c6e320f79, address scope support in subnetpool
INFO  [alembic.runtime.migration] Running upgrade 1b4c6e320f79 -> 48153cb5f051, qos db changes
INFO  [alembic.runtime.migration] Running upgrade 48153cb5f051 -> 9859ac9c136, quota_reservations
INFO  [alembic.runtime.migration] Running upgrade 9859ac9c136 -> 34af2b5c5a59, Add dns_name to Port
INFO  [alembic.runtime.migration] Running upgrade 34af2b5c5a59 -> 59cb5b6cf4d, Add availability zone
INFO  [alembic.runtime.migration] Running upgrade 59cb5b6cf4d -> 13cfb89f881a, add is_default to subnetpool
INFO  [alembic.runtime.migration] Running upgrade 13cfb89f881a -> 32e5974ada25, Add standard attribute table
INFO  [alembic.runtime.migration] Running upgrade 32e5974ada25 -> ec7fcfbf72ee, Add network availability zone
INFO  [alembic.runtime.migration] Running upgrade ec7fcfbf72ee -> dce3ec7a25c9, Add router availability zone
INFO  [alembic.runtime.migration] Running upgrade dce3ec7a25c9 -> c3a73f615e4, Add ip_version to AddressScope
INFO  [alembic.runtime.migration] Running upgrade c3a73f615e4 -> 659bf3d90664, Add tables and attributes to support external DNS integration
INFO  [alembic.runtime.migration] Running upgrade 659bf3d90664 -> 1df244e556f5, add_unique_ha_router_agent_port_bindings
INFO  [alembic.runtime.migration] Running upgrade 1df244e556f5 -> 19f26505c74f, Auto Allocated Topology - aka Get-Me-A-Network
INFO  [alembic.runtime.migration] Running upgrade 19f26505c74f -> 15be73214821, add dynamic routing model data
INFO  [alembic.runtime.migration] Running upgrade 15be73214821 -> b4caf27aae4, add_bgp_dragent_model_data
INFO  [alembic.runtime.migration] Running upgrade b4caf27aae4 -> 15e43b934f81, rbac_qos_policy
INFO  [alembic.runtime.migration] Running upgrade 15e43b934f81 -> 31ed664953e6, Add resource_versions row to agent table
INFO  [alembic.runtime.migration] Running upgrade 31ed664953e6 -> 2f9e956e7532, tag support
INFO  [alembic.runtime.migration] Running upgrade 2f9e956e7532 -> 3894bccad37f, add_timestamp_to_base_resources
INFO  [alembic.runtime.migration] Running upgrade 3894bccad37f -> 0e66c5227a8a, Add desc to standard attr table
INFO  [alembic.runtime.migration] Running upgrade kilo -> 30018084ec99, Initial no-op Liberty contract rule.
INFO  [alembic.runtime.migration] Running upgrade 30018084ec99, 8675309a5c4f -> 4ffceebfada, network_rbac
INFO  [alembic.runtime.migration] Running upgrade 4ffceebfada -> 5498d17be016, Drop legacy OVS and LB plugin tables
INFO  [alembic.runtime.migration] Running upgrade 5498d17be016 -> 2a16083502f3, Metaplugin removal
INFO  [alembic.runtime.migration] Running upgrade 2a16083502f3 -> 2e5352a0ad4d, Add missing foreign keys
INFO  [alembic.runtime.migration] Running upgrade 2e5352a0ad4d -> 11926bcfe72d, add geneve ml2 type driver
INFO  [alembic.runtime.migration] Running upgrade 11926bcfe72d -> 4af11ca47297, Drop cisco monolithic tables
INFO  [alembic.runtime.migration] Running upgrade 4af11ca47297 -> 1b294093239c, Drop embrane plugin table
INFO  [alembic.runtime.migration] Running upgrade 1b294093239c, 32e5974ada25 -> 8a6d8bdae39, standardattributes migration
INFO  [alembic.runtime.migration] Running upgrade 8a6d8bdae39 -> 2b4c2465d44b, DVR sheduling refactoring
INFO  [alembic.runtime.migration] Running upgrade 2b4c2465d44b -> e3278ee65050, Drop NEC plugin tables
INFO  [alembic.runtime.migration] Running upgrade e3278ee65050, 15e43b934f81 -> c6c112992c9, rbac_qos_policy
INFO  [alembic.runtime.migration] Running upgrade c6c112992c9 -> 5ffceebfada, network_rbac_external
INFO  [alembic.runtime.migration] Running upgrade 5ffceebfada, 0e66c5227a8a -> 4ffceebfcdc, standard_desc
  OK
```

注：进入 neutron 数据库查看是否有数据表，验证是否同步成功。

11. 启动 Neutron 服务并设置开机自启动

systemctl restart openstack-nova-api.service

systemctl enable neutron-server.service neutron-linuxbridge-agent.service neutron-dhcp-agent.service neutron-metadata-agent.serviceneutron-l3-agent.service

systemctl start neutron-server.service neutron-linuxbridge-agent.service neutron-dhcp-agent.service neutron-metadata-agent.serviceneutron-l3-agent.service

7.3 安装并配置计算节点

1. 安装 Neutron 组件所需软件包

yum install openstack-neutron-linuxbridge ebtables ipset -y

2. 配置 Neutron 所需组件

编辑/etc/neutron/neutron.conf 文件。

编辑[DEFAULT]和[oslo_messaging_rabbit]部分，配置 RabbitMQ 消息服务器链接。

[DEFAULT]
rpc_backend = rabbit

[oslo_messaging_rabbit]
rabbit_host = controller
rabbit_userid = openstack
rabbit_password = 000000

编辑[DEFAULT]和[keystone_authtoken]部分，配置 Keystone 身份认证。

[DEFAULT]
auth_strategy = keystone

[keystone_authtoken]
auth_uri = http://controller:5000
auth_url = http://controller:35357
memcached_servers = controller:11211
auth_type = password
project_domain_name = default
user_domain_name = default
project_name = service
username = neutron
password = 000000   #创建 neutron 用户设置的密码，自定义

编辑[oslo_concurrency]部分，配置 loca_path。

[oslo_concurrency]
lock_path = /var/lib/neutron/tmp

3. 配置 linux bridge 插件

编辑/etc/neutron/plugins/ml2/linuxbridge_agent.ini 文件。

编辑[linux_bridge]部分，配置虚拟网络映射到物理网络接口。

[linux_bridge]
physical_interface_mappings = provider: enp1s0f1 （物理机的外网网卡名）

编辑[vxlan]部分，使 VxLAN 覆盖网络，并配置物理网络的 IP 地址。

```
[vxlan]
enable_vxlan = True
local_ip = 192.168.100.20
l2_population = True
```

编辑[securitygroup]部分，配置安全组和网桥配置防火墙驱动。

```
[securitygroup]
enable_security_group = True
firewall_driver = neutron.agent.linux.iptables_firewall.IptablesFirewallDriver
```

4. 配置 Nova 服务使用网络

编辑/etc/nova/nova.conf 文件。

编辑[neutron]部分，配置访问参数。

```
[neutron]
url = http://controller:9696
auth_url = http://controller:35357
auth_type = password
project_domain_name = default
user_domain_name = default
region_name = RegionOne
project_name = service
username = neutron
password = 000000          #创建 neutron 用户的密码
```

5. 启动 Neutron 服务并设置开机自启动

```
# systemctl restart openstack-nova-compute.service
# systemctl enable neutron-linuxbridge-agent.service
# systemctl start neutron-linuxbridge-agent.service
```

7.4 验证 Neutron 服务

1. 控制节点生效 admin 用户环境变量

```
# . admin-openrc
```

2. 查看 Neutron 服务

```
# neutron ext-list
```

```
[root@controller ~]# neutron ext-list
+---------------------------------+---------------------------------------------+
| alias                           | name                                        |
+---------------------------------+---------------------------------------------+
| default-subnetpools             | Default Subnetpools                         |
| network-ip-availability         | Network IP Availability                     |
| network_availability_zone       | Network Availability Zone                   |
| auto-allocated-topology         | Auto Allocated Topology Services            |
| ext-gw-mode                     | Neutron L3 Configurable external gateway mode |
| binding                         | Port Binding                                |
| agent                           | agent                                       |
| subnet_allocation               | Subnet Allocation                           |
| l3_agent_scheduler              | L3 Agent Scheduler                          |
| tag                             | Tag support                                 |
| external-net                    | Neutron external network                    |
| net-mtu                         | Network MTU                                 |
| availability_zone               | Availability Zone                           |
| quotas                          | Quota management support                    |
| l3-ha                           | HA Router extension                         |
| provider                        | Provider Network                            |
| multi-provider                  | Multi Provider Network                      |
| address-scope                   | Address scope                               |
| extraroute                      | Neutron Extra Route                         |
| timestamp_core                  | Time Stamp Fields addition for core resources |
| router                          | Neutron L3 Router                           |
| extra_dhcp_opt                  | Neutron Extra DHCP opts                     |
| dns-integration                 | DNS Integration                             |
| security-group                  | security-group                              |
| dhcp_agent_scheduler            | DHCP Agent Scheduler                        |
| router_availability_zone        | Router Availability Zone                    |
| rbac-policies                   | RBAC Policies                               |
| standard-attr-description       | standard-attr-description                   |
| port-security                   | Port Security                               |
| allowed-address-pairs           | Allowed Address Pairs                       |
| dvr                             | Distributed Virtual Router                  |
+---------------------------------+---------------------------------------------+
```

neutron agent-list

```
[root@controller ~]# neutron agent-list
+------------+------------+----------+-------------------+-------+----------------+
| id         | agent_type | host     | availability_zone | alive | admin_state_up |
| binary     |            |          |                   |       |                |
+------------+------------+----------+-------------------+-------+----------------+
| 09495dd8-  | Linux      | compute  |                   | :-)   | True           |
| neutron-li |            |          |                   |       |                |
| c2f8-4244  | bridge     |          |                   |       |                |
| nuxbridge- |            |          |                   |       |                |
| -884f-942  | agent      |          |                   |       |                |
| agent      |            |          |                   |       |                |
| 54f40d61c  |            |          |                   |       |                |
|            |            |          |                   |       |                |
| 33da13ee-  | Linux      | controlle|                   | :-)   | True           |
| neutron-li |            |          |                   |       |                |
| ff2c-4338  | bridge     | r        |                   |       |                |
| nuxbridge- |            |          |                   |       |                |
| -bfa3-789  | agent      |          |                   |       |                |
| agent      |            |          |                   |       |                |
| 09e73ba2b  |            |          |                   |       |                |
|            |            |          |                   |       |                |
| 5b4ecfc8-  | DHCP agent | controlle| nova              | :-)   | True           |
| neutron-   |            |          |                   |       |                |
| 8134-4631  |            | r        |                   |       |                |
| dhcp-agent |            |          |                   |       |                |
| -ad9b-58b  |            |          |                   |       |                |
|            |            |          |                   |       |                |
| 3667b5a1c  |            |          |                   |       |                |
|            |            |          |                   |       |                |
| cba709bf-  | L3 agent   | controlle| nova              | :-)   | True           |
| neutron-l3 |            |          |                   |       |                |
| bc29-41a7  |            | r        |                   |       |                |
| -agent     |            |          |                   |       |                |
| -8821-e82  |            |          |                   |       |                |
|            |            |          |                   |       |                |
| dc59975e2  |            |          |                   |       |                |
|            |            |          |                   |       |                |
| d205f65d-  | Metadata   | controlle|                   | :-)   | True           |
| neutron-   |            |          |                   |       |                |
| 46f8       | agent      | r        |                   |       |                |
| metadata-  |            |          |                   |       |                |
| -4e7e-8f4  |            |          |                   |       |                |
| agent      |            |          |                   |       |                |
| 8-6a03fe8  |            |          |                   |       |                |
|            |            |          |                   |       |                |
| 50b0c      |            |          |                   |       |                |
+------------+------------+----------+-------------------+-------+----------------+
```

注：Linux bridge agent，Linux bridge agent，DHCP agent，L3 agent，Metadata agent 都要列出来，即服务正常。

课后习题

1. 同步 neutron 数据库的命令为_____。
2. 在 Neutron 组件中，Neutron 服务的内部端点地址为_____。
3. 为/etc/neutron/plugin.ini 文件创建软链接的命令为_____。
4. 如何查看 Neutron 服务状态？

第 8 章 对象存储服务 Swift

8.1 Swift 基本概念

OpenStack Object Storage（Swift）是 OpenStack 开源云计算项目的子项目之一。Swift 的目的是使用普通硬件来构建冗余的、可扩展的分布式对象存储集群，存储容量可达 PB 级。

Swift 并不是文件系统或者实时的数据存储系统，它是对象存储，用于永久类型的静态数据的长期存储，对这些数据可以检索、调整，必要时进行更新。最适合存储的数据类型的例子是虚拟机镜像、图片存储、邮件存储和存档备份。

Swift 无须采用 RAID（磁盘冗余阵列），也没有中心单元或主控节点。

Swift 通过在软件层面引入一致性哈希技术和数据冗余性，牺牲一定程度的数据一致性来达到高可用性（High Availability，HA）和可伸缩性，支持多租户模式、容器和对象读写操作，适合解决互联网的应用场景下非结构化数据存储问题。

注：对象存储服务不使用 SQL 数据库，而是在每一个对象存储节点上使用分布式的 SQLite 数据库。在安装之前，存储节点应准备两个空本地存储设备，本次实验为 **/dev/sdb** 和 **/dev/sdc**。

8.2 控制节点环境配置

为对象存储服务创建服务认证和 API 端点。

1. 生效 admin 用户环境变量

```
# . admin-openrc
```

2. 创建 swift 用户

```
# openstack user create --domain default --password-prompt swift
```

```
[root@controller ~]# openstack user create --domain default --password-prompt sw
ift
User Password:
Repeat User Password:
+-----------+------------------------------------+
| Field     | Value                              |
+-----------+------------------------------------+
| domain_id | a156af4f59c449ca88f767e40c99b691   |
| enabled   | True                               |
| id        | bed9d8d7d8ba4ad3bd1bd46199e059e4   |
| name      | swift                              |
+-----------+------------------------------------+
```

3. 将 admin 角色添加给 swift 用户

 # openstack role add --project service --user swift admin

```
[root@controller ~]# openstack role add --project service --user swift admin
[root@controller ~]#
```

4. 创建 swift 服务实体

 # openstack service create --name swift --description "OpenStack Object Storage" object-store

```
[root@controller ~]# openstack service create --name swift  --description "OpenStack Object
Storage" object-store
+-------------+----------------------------------+
| Field       | Value                            |
+-------------+----------------------------------+
| description | OpenStack Object Storage         |
| enabled     | True                             |
| id          | bf65073dcaaa4829b83d5898af9c74e3 |
| name        | swift                            |
| type        | object-store                     |
+-------------+----------------------------------+
```

5. 创建 swift 服务 API 端点

创建公共端点：

 # openstack endpoint create --region RegionOne object-store public http://controller:8080/v1/AUTH_%\(tenant_id\)s

创建外部端点：

 # openstack endpoint create --region RegionOne object-store internal http://controller:8080/v1/AUTH_%\(tenant_id\)s

创建管理端点：

 # openstack endpoint create --region RegionOne object-store admin http://controller:8080/v1

```
[root@controller ~]# openstack endpoint create --region RegionOne object-store public http:
//controller:8080/v1/AUTH_%\(tenant_id\)s
+--------------+-----------------------------------------+
| Field        | Value                                   |
+--------------+-----------------------------------------+
| enabled      | True                                    |
| id           | ea52f827e44c4facbd1011deacc2a9f5        |
| interface    | public                                  |
| region       | RegionOne                               |
| region_id    | RegionOne                               |
| service_id   | bf65073dcaaa4829b83d5898af9c74e3        |
| service_name | swift                                   |
| service_type | object-store                            |
| url          | http://controller:8080/v1/AUTH_%(tenant_id)s |
+--------------+-----------------------------------------+
[root@controller ~]# openstack endpoint create --region RegionOne  object-store internal ht
tp://controller:8080/v1/AUTH_%\(tenant_id\)s
+--------------+-----------------------------------------+
| Field        | Value                                   |
+--------------+-----------------------------------------+
```

```
| enabled      | True                                              |
| id           | 26c821fb7bc049808c78c46cab9b98af                  |
| interface    | internal                                          |
| region       | RegionOne                                         |
| region_id    | RegionOne                                         |
| service_id   | bf65073dcaaa4829b83d5898af9c74e3                  |
| service_name | swift                                             |
| service_type | object-store                                      |
| url          | http://controller:8080/v1/AUTH_%(tenant_id)s      |

[root@controller ~]# openstack  endpoint  create --region RegionOne object-store admin ht
tp://controller:8080/v1
+--------------+--------------------------------------+
| Field        | Value                                |
+--------------+--------------------------------------+
| enabled      | True                                 |
| id           | 691ab3ef446b4e5aac29ad554ec484c8     |
| interface    | admin                                |
| region       | RegionOne                            |
| region_id    | RegionOne                            |
| service_id   | bf65073dcaaa4829b83d5898af9c74e3     |
| service_name | swift                                |
| service_type | object-store                         |
| url          | http://controller:8080/v1            |
+--------------+--------------------------------------+
[root@controller ~]#
```

8.3 控制节点安装并配置 Swift

1. 安装 Swift 软件包

　　# yum install openstack-swift-proxy python-swiftclient python-keystoneclient python-keystonemiddleware memcached -y

2. 从 OpenStack 官网对象存储仓库源中获取代理服务的配置文件

　　# curl –o /etc/swift/proxy-server.conf https://git.openstack.org/cgit/openstack/swift/plain/etc/proxy-server.conf-sample?h=stable/mitaka

　　注：这里需要将我们的设备连接外网，如果没有网络，可用浏览器直接打开这个链接，将里面的内容复制到 proxy-server.conf 这个文件内，请注意文件路径。

3. 编辑并配置文件 /etc/swift/proxy-server.conf

　　在[DEFAULT]部分，配置绑定端口，用户和配置文件放置的目录。

[DEFAULT]

bind_port = 8080
swift_dir = /etc/swift
user = swift

　　在[pipeline:main]部分，删除"tempurl"和"tempauth"模块，并增加"authtoken"和"keystoneauth"模块。

[pipeline:main]

pipeline = catch_errors gatekeeper healthcheck proxy-logging cache container_sync bulk ratelimit authtoken keystoneauth container-quotas account-quotas slo dlo versioned_writes proxy-logging proxy-server

　　在[app:proxy-server]部分，启动自动创建账户。

```
[app:proxy-server]
```

use = egg:swift#proxy
account_autocreate = True

在[filter:keystoneauth]部分，配置 operator 的角色。

```
[filter:keystoneauth]
```

use = egg:swift#keystoneauth
operator_roles = admin,user

在[filter:authtoken]部分，配置身份认证服务的访问。

```
[filter:authtoken]
```

paste.filter_factory = keystonemiddleware.auth_token:filter_factory
auth_uri = http://controller:5000
auth_url = http://controller:35357
memcached_servers = controller:11211
auth_type = password
project_domain_name = default
user_domain_name = default
project_name = service
username = swift
password = 000000
delay_auth_decision = True

在[filter:cache]部分，配置 memcache 的地址。

```
[filter:cache]
```

use = egg:swift#memcache
memcache_servers = 127.0.0.1:11211

8.4 存储节点安装并配置 Swift

本节描述怎样为操作账号、容器和对象服务安装和配置存储节点。

为简单起见，这里配置两个存储节点，每个节点包含两个空本地块存储设备**/dev/sdb** 和 **/dev/sdc**。

注：因为有两个存储节点，所以需要按照本书的第 3 章、第 4 章、第 6 章、第 7 章做准备环境的操作。

（1）存储节点 1 的管理 IP 地址为 192.168.100.40，主机名为 storage2。

存储节点 2 的管理 IP 地址为 192.168.100.50，主机名为 storage3。

（2）配置 hosts 文件配对，对所有的主机都需要进行更新配置。

（3）安全配置（关闭防火墙、Selinux）、配置 yum 源、NTP 以及安装 OpenStack 包。

8.4.1 环境准备

以下操作需要在两个存储节点上分别执行。

1. 安装支持的工具包

```
# yum install xfsprogs rsync -y
```

2. 使用 XFS 格式化 "/dev/sdb" 和 "/dev/sdc" 两块磁盘

```
# mkfs.xfs /dev/sdb
# mkfs.xfs /dev/sdc
```

```
[root@storage2 ~]# mkfs.xfs /dev/sdb
meta-data=/dev/sdb               isize=256    agcount=4, agsize=655360 blks
         =                       sectsz=512   attr=2, projid32bit=1
         =                       crc=0        finobt=0
data     =                       bsize=4096   blocks=2621440, imaxpct=25
         =                       sunit=0      swidth=0 blks
naming   =version 2               bsize=4096   ascii-ci=0 ftype=0
log      =internal log            bsize=4096   blocks=2560, version=2
         =                       sectsz=512   sunit=0 blks, lazy-count=1
realtime =none                   extsz=4096   blocks=0, rtextents=0
[root@storage2 ~]# mkfs.xfs /dev/sdc
meta-data=/dev/sdc               isize=256    agcount=4, agsize=655360 blks
         =                       sectsz=512   attr=2, projid32bit=1
         =                       crc=0        finobt=0
data     =                       bsize=4096   blocks=2621440, imaxpct=25
         =                       sunit=0      swidth=0 blks
naming   =version 2               bsize=4096   ascii-ci=0 ftype=0
log      =internal log            bsize=4096   blocks=2560, version=2
         =                       sectsz=512   sunit=0 blks, lazy-count=1
realtime =none                   extsz=4096   blocks=0, rtextents=0
[root@storage2 ~]#
```

3. 创建磁盘挂载目录

```
# mkdir -p /srv/node/sdb
# mkdir -p /srv/node/sdc
```

4. 编辑/etc/fstab 文件，添加如下文件配置自动开机自动挂载

```
/dev/sdb /srv/node/sdb xfs noatime,nodiratime,nobarrier,logbufs=8 0 2
/dev/sdc /srv/node/sdc xfs noatime,nodiratime,nobarrier,logbufs=8 0 2
```

5. 挂载磁盘

```
# mount /srv/node/sdb
# mount /srv/node/sdc
# df  -hT      （查看是否挂载成功）
```

```
[root@storage2 ~]# df -hT
Filesystem              Type     Size  Used Avail Use% Mounted on
/dev/mapper/centos-root xfs       19G   12G  7.5G  61% /
devtmpfs                devtmpfs 480M     0  480M   0% /dev
tmpfs                   tmpfs    489M     0  489M   0% /dev/shm
tmpfs                   tmpfs    489M  6.7M  483M   2% /run
tmpfs                   tmpfs    489M     0  489M   0% /sys/fs/cgroup
/dev/sda1               xfs      187M  150M   38M  81% /boot
tmpfs                   tmpfs     98M     0   98M   0% /run/user/0
/dev/sdb                xfs       10G   33M   10G   1% /srv/node/sdb
/dev/sdc                xfs       10G   33M   10G   1% /srv/node/sdc
```

6. 编辑/etc/rsyncd.conf 文件，包含以下内容

```
uid = swift
gid = swift
```

```
log file = /var/log/rsyncd.log
pid file = /var/run/rsyncd.pid
address = MANAGEMENT_INTERFACE_IP_ADDRESS（存储节点管理 IP）

[account]
max connections = 2
path = /srv/node/
read only = False
lock file = /var/lock/account.lock

[container]
max connections = 2
path = /srv/node/
read only = False
lock file = /var/lock/container.lock

[object]
max connections = 2
path = /srv/node/
read only = False
lock file = /var/lock/object.lock
```

7. 启动 rsyncd 服务并设置开机自启动

```
# systemctl enable rsyncd.service
# systemctl start rsyncd.service
```

8.4.2 安装并配置

注：以下操作需要在每一个存储节点上执行。

1. 安装软件包

```
# yum install openstack-swift-account openstack-swift-container openstack-swift-object -y
```

2. 从 OpenStack 官网对象存储源仓库中获取 accounting、container 以及 object 服务配置文件

```
# curl -o /etc/swift/account-server.conf https://git.openstack.org/cgit/openstack/swift/plain/etc/account-server.conf-sample?h=stable/mitaka
# curl -o /etc/swift/container-server.conf https://git.openstack.org/cgit/openstack/swift/plain/etc/container-server.conf-sample?h=stable/mitaka
# curl -o /etc/swift/object-server.conf https://git.openstack.org/cgit/openstack/swift/plain/etc/object-server.conf-sample?h=stable/mitaka
```

3. 编辑/etc/swift/account-server.conf 文件并完成下列操作

编辑[DEFAULT]部分，配置绑定 IP 地址，绑定端口、用户，配置目录和挂载目录。

```
[DEFAULT]

bind_ip = MANAGEMENT_INTERFACE_IP_ADDRESS（存储节点管理 IP）
```

```
bind_port = 6002
# bind_timeout = 30
# backlog = 4096
user = swift
swift_dir = /etc/swift
devices = /srv/node
mount_check = True
```

编辑**[pipeline:main]**部分，启用合适的模块。

```
[pipeline:main]
pipeline = healthcheck recon account-server
```

编辑**[filter:recon]**部分，配置 recon（meters）缓存目录。

```
[filter:recon]
use = egg:swift#recon
recon_cache_path = /var/cache/swift
```

4. 编辑/etc/swift/container-server.conf 文件并完成下列操作

编辑**[DEFAULT]**部分，配置绑定 IP 地址，绑定端口、用户，配置目录和挂载目录。

```
[DEFAULT]

bind_ip = MANAGEMENT_INTERFACE_IP_ADDRESS（存储节点管理 IP）
bind_port = 6001
user = swift
swift_dir = /etc/swift
devices = /srv/node
mount_check = True
```

编辑**[pipeline:main]**部分，启用合适的模块。

```
[pipeline:main]

pipeline = healthcheck recon container-server
```

编辑**[filter:recon]**部分，配置 recon（meters）缓存目录。

```
[filter:recon]
use = egg:swift#recon
recon_cache_path = /var/cache/swift
```

5. 编辑/etc/swift/object-server.conf 文件并完成下列操作

编辑**[DEFAULT]**部分，配置绑定 IP 地址，绑定端口、用户，配置目录和挂载目录。

```
[DEFAULT]

bind_ip = MANAGEMENT_INTERFACE_IP_ADDRESS（存储节点管理 IP）
bind_port = 6000
user = swift
swift_dir = /etc/swift
```

devices = /srv/node
mount_check = True

编辑[pipeline:main]部分，启用合适的模块。

[pipeline:main]

pipeline = healthcheck recon object-server

编辑[filter:recon]部分，配置 recon（meters）缓存目录。

[filter:recon]

use = egg:swift#recon
recon_cache_path = /var/cache/swift
recon_lock_path = /var/lock

6. 修改挂载点目录权限

chown -R swift:swift /srv/node

查看：

ls -l /srv/

```
[root@storage2 ~]# ls -l /srv/
total 0
drwxr-xr-x. 4 swift_swift 26 Dec 13 11:51 node
```

7. 创建 recon 的目录并修改权限

mkdir -p /var/cache/swift
chown -R root:swift /var/cache/swift
chmod -R 775 /var/cache/swift

8.5 创建并分发 Ring

以下操作在控制节点上执行。

8.5.1 创建账户 Ring

1. 切换到/etc/swift 目录

创建基本 account.builder 文件：

swift-ring-builder account.builder create 10 2 1

2. 添加所有存储节点到 Ring 中

account.builder add --region 1 -zone 1 --ip STORAGE_NODE_MANAGEMENT_ INTERFACE_ IP_ADDRESS --port 6002 --device DEVICE_NAME --weight DEVICE_WEIGHT

说明：将 STORAGE_NODE_MANAGEMENT_INTERFACE_IP_ADDRESS 替换为存储节点管理网络的 IP 地址。将 DEVICE_NAME 替换为同一个存储节点存储设备名称。

DEVICE_WEIGHT 为大小。

```
# swift-ring-builder account.builder add --region 1 --zone 1 --ip 192.168.100.40 --port 6002 --device sdb --weight 100

# swift-ring-builder account.builder add --region 1 --zone 1 --ip 192.168.100.40 --port 6002 --device sdc --weight 100

# swift-ring-builder account.builder add --region 1 --zone 2 --ip 192.168.100.50 --port 6002 --device sdb --weight 100

# swift-ring-builder account.builder add --region 1 --zone 2 --ip 192.168.100.50 --port 6002 --device sdc --weight 100
```

3. 校验 Ring 的内容

```
# swift-ring-builder account.builder
```

```
[root@controller swift]# swift-ring-builder account.builder
account.builder, build version 4
1024 partitions, 2.000000 replicas, 1 regions, 2 zones, 4 devices, 100.00 balance, 0.00 dispersion
The minimum number of hours before a partition can be reassigned is 1 (0:00:00 remaining)
The overload factor is 0.00% (0.000000)
Ring file account.ring.gz not found, probably it hasn't been written yet
Devices:    id  region  zone      ip address  port  replication ip  replication port      name weight partitions balance flags meta
             0       1     1  192.168.100.40  6002  192.168.100.40            6002      sdb 100.00      0  -100.00
             1       1     1  192.168.100.40  6002  192.168.100.40            6002      sdc 100.00      0  -100.00
             2       1     2  192.168.100.50  6002  192.168.100.50            6002      sdb 100.00      0  -100.00
             3       1     2  192.168.100.50  6002  192.168.100.50            6002      sdc 100.00      0  -100.00
```

4. 重新分发 Ring

```
# swift-ring-builder account.builder rebalance
```

```
[root@controller swift]# swift-ring-builder account.builder rebalance
Reassigned 2048 (200.00%) partitions. Balance is now 0.00. Dispersion is now 0.00
```

8.5.2 创建容器 Ring

1. 切换到 /etc/swift 目录

创建基本 container.builder 文件。

```
[root@controller swift]# cd /etc/swift/
[root@controller swift]# swift-ring-builder account.builder create 10 2 1
[root@controller swift]# swift-ring-builder account.builder add --region 1 --zone 1 --ip 192.168.100.40 --port 6002 --device sdb --weight 100
Device d0r1z1-192.168.100.40:6002R192.168.100.40:6002/sdb_"" with 100.0 weight got id 0
[root@controller swift]# swift-ring-builder account.builder add --region 1 --zone 1 --ip 192.168.100.40 --port 6002 --device sdc --weight 100
Device d1r1z1-192.168.100.40:6002R192.168.100.40:6002/sdc_"" with 100.0 weight got id 1
[root@controller swift]# swift-ring-builder account.builder add --region 1 --zone 2 --ip 192.168.100.50 --port 6002 --device sdb --weight 100
Device d2r1z2-192.168.100.50:6002R192.168.100.50:6002/sdb_"" with 100.0 weight got id 2
[root@controller swift]# swift-ring-builder account.builder add --region 1 --zone 2 --ip 192.168.100.50 --port 6002 --device sdc --weight 100
Device d3r1z2-192.168.100.50:6002R192.168.100.50:6002/sdc_"" with 100.0 weight got id 3
[root@controller swift]#
```

```
# swift-ring-builder container.builder create 10 2 1
```

2. 添加所有存储节点到 Ring 中

```
# swift-ring-builder container.builder add --region 1 --zone 1 --ip 192.168.100.40 --port 6001 --device sdb --weight 100
```

    # swift-ring-builder container.builder add　--region 1　--zone 1　--ip 192.168.100.40　--port 6001 --device sdc --weight 100

    # swift-ring-builder container.builder add　--region 1　--zone 2　--ip 192.168.100.50　--port 6001 --device sdb --weight 100

    # swift-ring-builder container.builder add　--region 1　--zone 2 --ip 192.168.100.50　--port 6001 --device sdc --weight 100

3. 校验 Ring 的内容

    # swift-ring-builder container.builder

```
[root@controller swift]# swift-ring-builder container.builder
container.builder, build version 4
1024 partitions, 2.000000 replicas, 1 regions, 2 zones, 4 devices, 100.00 balance, 0.00 dispersion
The minimum number of hours before a partition can be reassigned is 1 (0:00:00 remaining)
The overload factor is 0.00% (0.000000)
Ring file container.ring.gz not found, probably it hasn't been written yet
Devices:    id region zone      ip address  port  replication ip  replication port      name weight partitions balance flags meta
             0      1    1  192.168.100.40  6001  192.168.100.40            6001      sdb 100.00          0 -100.00
             1      1    1  192.168.100.40  6001  192.168.100.40            6001      sdc 100.00          0 -100.00
             2      1    2  192.168.100.50  6001  192.168.100.50            6001      sdb 100.00          0 -100.00
             3      1    2  192.168.100.50  6001  192.168.100.50            6001      sdc 100.00          0 -100.00
[root@controller swift]#
[root@controller swift]# swift-ring-builder container.builder create 10 2 1
[root@controller swift]# swift-ring-builder container.builder add    --region 1 --zone 1 --ip 192.168.100.40 --port 6001 --device sdb --weight 100
Device d0r1z1-192.168.100.40:6001R192.168.100.40:6001/sdb_"" with 100.0 weight got id 0
[root@controller swift]#
[root@controller swift]# swift-ring-builder container.builder add    --region 1 --zone 1 --ip 192.168.100.40 --port 6001 --device sdc --weight 100
Device d1r1z1-192.168.100.40:6001R192.168.100.40:6001/sdc_"" with 100.0 weight got id 1
[root@controller swift]#
[root@controller swift]# swift-ring-builder container.builder add    --region 1 --zone 2 --ip 192.168.100.50 --port 6001 --device sdb --weight 100
Device d2r1z2-192.168.100.50:6001R192.168.100.50:6001/sdb_"" with 100.0 weight got id 2
[root@controller swift]# swift-ring-builder container.builder add    --region 1 --zone 2 --ip 192.168.100.50 --port 6001 --device sdc --weight 100
Device d3r1z2-192.168.100.50:6001R192.168.100.50:6001/sdc_"" with 100.0 weight got id 3
[root@controller swift]#
```

4. 重新分发 Ring

    # swift-ring-builder container.builder rebalance

```
[root@controller swift]# swift-ring-builder container.builder rebalance
Reassigned 2048 (200.00%) partitions. Balance is now 0.00.   Dispersion is now 0.00
```

8.5.3 创建对象 Ring

1. 切换到/etc/swift 目录

创建基本 object.builder 文件：

 # swift-ring-builder object.builder create 10 2 1

2. 添加所有存储节点到 Ring 中

    # swift-ring-builder object.builder add　--region 1 --zone 1 --ip 192.168.100.40 --port 6000 --device sdb --weight 100

```
# swift-ring-builder object.builder add    --region 1 --zone 1 --ip 192.168.100.40 --port 6000 --device sdc --weight 100
# swift-ring-builder object.builder add    --region 1 --zone 2 --ip 192.168.100.50 --port 6000 --device sdb --weight 100
# swift-ring-builder object.builder add    --region 1 --zone 2 --ip 192.168.100.50 --port 6000 --device sdc --weight 100
```

```
[root@controller swift]# swift-ring-builder object.builder create 10 2 1
[root@controller swift]#
[root@controller swift]# swift-ring-builder object.builder add    --region 1 --zone 1 --ip 192.168.100.40 --port 6000 --device sdb --weight 100
Device d0r1z1-192.168.100.40:6000R192.168.100.40:6000/sdb_"" with 100.0 weight got id 0
[root@controller swift]# swift-ring-builder object.builder add    --region 1 --zone 1 --ip 192.168.100.40 --port 6000 --device sdc --weight 100
Device d1r1z1-192.168.100.40:6000R192.168.100.40:6000/sdc_"" with 100.0 weight got id 1
[root@controller swift]# swift-ring-builder object.builder add    --region 1 --zone 2 --ip 192.168.100.50 --port 6000 --device sdb --weight 100
Device d2r1z2-192.168.100.50:6000R192.168.100.50:6000/sdb_"" with 100.0 weight got id 2
[root@controller swift]# swift-ring-builder object.builder add    --region 1 --zone 2 --ip 192.168.100.50 --port 6000 --device sdc --weight 100
Device d3r1z2-192.168.100.50:6000R192.168.100.50:6000/sdc_"" with 100.0 weight got id 3
[root@controller swift]#
```

3. 校验 Ring 的内容

```
# swift-ring-builder object.builder
```

```
[root@controller swift]# swift-ring-builder object.builder
object.builder, build version 4
1024 partitions, 2.000000 replicas, 1 regions, 2 zones, 4 devices, 100.00 balance, 0.00 dispersion
The minimum number of hours before a partition can be reassigned is 1 (0:00:00 remaining)
The overload factor is 0.00% (0.000000)
Ring file object.ring.gz not found, probably it hasn't been written yet
Devices:    id  region  zone      ip address  port  replication ip  replication port      name weight partitions balance flags meta
             0       1     1  192.168.100.40  6000  192.168.100.40            6000       sdb 100.00          0 -100.00
             1       1     1  192.168.100.40  6000  192.168.100.40            6000       sdc 100.00          0 -100.00
             2       1     2  192.168.100.50  6000  192.168.100.50            6000       sdb 100.00          0 -100.00
             3       1     2  192.168.100.50  6000  192.168.100.50            6000       sdc 100.00          0 -100.00
[root@controller swift]#
```

4. 重新分发 Ring

```
# swift-ring-builder object.builder rebalance
```

```
[root@controller swift]# swift-ring-builder object.builder rebalance
Reassigned 2048 (200.00%) partitions. Balance is now 0.00. Dispersion is now 0.00
```

8.5.4 完成安装

（1）控制节点分化 Ring 文件到每一个存储节点，复制 account.ring.gz、container.ring.gz、object.ring.gz 文件到每个存储节点的/etc/swift 目录中。

```
# scp   *.gz   192.168.100.40:/etc/swift/
# scp   *.gz   192.168.100.50:/etc/swift/
```

```
[root@controller swift]# scp *.gz 192.168.100.40:/etc/swift/
The authenticity of host '192.168.100.40 (192.168.100.40)' can't be established.
ECDSA key fingerprint is e0:49:2f:03:78:52:cb:37:77:93:7b:07:68:8e:bd:1d.
Are you sure you want to continue connecting (yes/no)? yes
Warning: Permanently added '192.168.100.40' (ECDSA) to the list of known hosts.
root@192.168.100.40's password:
account.ring.gz                                    100%  708     0.7KB/s   00:00
container.ring.gz                                  100%  721     0.7KB/s   00:00
object.ring.gz                                     100%  707     0.7KB/s   00:00
```

```
[root@controller swift]# scp *.gz 192.168.100.50:/etc/swift/
The authenticity of host '192.168.100.50 (192.168.100.50)' can't be established.
ECDSA key fingerprint is e0:49:2f:03:78:52:cb:37:77:93:7b:07:68:8e:bd:1d.
Are you sure you want to continue connecting (yes/no)? yes
Warning: Permanently added '192.168.100.50' (ECDSA) to the list of known hosts.
root@192.168.100.50's password:
account.ring.gz                              100%  708     0.7KB/s   00:00
container.ring.gz                            100%  721     0.7KB/s   00:00
object.ring.gz                               100%  707     0.7KB/s   00:00
[root@controller swift]#
```

（2）控制节点从 OpenStack 对象存储源仓库中获取/etc/swift/swift.conf 文件。

curl -o /etc/swift/swift.conf \
    https://git.openstack.org/cgit/openstack/swift/plain/etc/swift.conf-sample?h=stable/mitaka

（3）编辑**/etc/swift/swift.conf** 文件。

在**[swift-hash]**部分，为环境配置路径前缀和后缀：

[swift-hash]

swift_hash_path_suffix = HASH_PATH_SUFFIX
swift_hash_path_prefix = HASH_PATH_PREFIX

将其中的 HASH_PATH_PREFIX 和 HASH_PATH_SUFFIX 替换为唯一的值。

在**[storage-policy:0]**部分，配置默认存储策略：

[storage-policy:0]

name = Policy-0
default = yes

（4）复制 swift.conf 文件到每个存储节点的/etc/swift 目录。

scp swift.conf 192.168.100.40:/etc/swift/
scp swift.conf 192.168.100.50:/etc/swift/

```
[root@controller swift]# scp swift.conf 192.168.100.40:/etc/swift/
root@192.168.100.40's password:
swift.conf                                   100% 7555     7.4KB/s   00:00
[root@controller swift]# scp swift.conf 192.168.100.50:/etc/swift/
root@192.168.100.50's password:
swift.conf                                   100% 7555     7.4KB/s   00:00
```

（5）在控制节点上，启动对象存储代理服务及其依赖服务，并配置开机启动。

systemctl enable openstack-swift-proxy.service memcached.service
systemctl start openstack-swift-proxy.service memcached.service

（6）在所有的存储节点上，启动对象存储服务，并将其设置开机启动。

  # systemctl enable openstack-swift-account.service openstack-swift-account-auditor.service openstack-swift-account-reaper.service openstack-swift-account-replicator.service
    # systemctl start openstack-swift-account.service openstack-swift-account-auditor.service openstack-swift-account-reaper.service openstack-swift-account-replicator.service
    # systemctl enable openstack-swift-container.service openstack-swift-container-auditor.service openstack-swift-container-replicator.service openstack-swift-container-updater.service
    # systemctl start openstack-swift-container.service openstack-swift-container-auditor.service openstack-swift-container-replicator.service openstack-swift-container-updater.service
    # systemctl enable openstack-swift-object.service openstack-swift-object-auditor.service openstack-swift-object-replicator.service openstack-swift-object-updater.service

systemctl start openstack-swift-object.service openstack-swift-object-auditor.service openstack-swift-object-replicator.service openstack-swift-object-updater.service

8.6 校验安装

在控制节点进行如下操作。

1. 生效 admin 用户环境变量

. /root/admin-openrc

2. 查看服务状态

swift stat

```
[root@controller swift]# . /root/admin-openrc
[root@controller swift]# swift stat
       Account: AUTH_1ac8bf3783e641e7b20f81b6e0f02dc1
    Containers: 0
       Objects: 0
         Bytes: 0
X-Put-Timestamp: 1481660158.76364
   X-Timestamp: 1481660158.76364
    X-Trans-Id: tx248ee347ffdd407784d5c-00585056fd
  Content-Type: text/plain; charset=utf-8
[root@controller swift]#
```

3. 创建名为 container1 的容器

openstack container create container1

```
[root@controller swift]# openstack container create container1
+----------------------------------+------------+------------------------------------+
| account                          | container  | x-trans-id                         |
+----------------------------------+------------+------------------------------------+
| AUTH_1ac8bf3783e641e7b20f81b6e0f02dc1 | container1 | tx55f9444385ef443198e51-005850584c |
+----------------------------------+------------+------------------------------------+
[root@controller swift]#
```

4. 上传一个测试文件（名为 FILE）到 container1 容器

openstack object create container1 FILE

```
[root@controller swift]# openstack object create container1 FILE
+--------+------------+----------------------------------+
| object | container  | etag                             |
+--------+------------+----------------------------------+
| FILE   | container1 | 4a64b2c810525316062f95747e3cdfa2 |
+--------+------------+----------------------------------+
[root@controller swift]#
```

5. 列出 container1 容器里的所有文件

openstack object list container1

```
[root@controller swift]# openstack object list container1
+------+
| Name |
+------+
| FILE |
+------+
```

6. 从 container1 容器里下载一个文件（FILE）

openstack object save container1 FILE

Swift 和 Cinder 的比较如下。

（1）Swift 是 Object Storage（对象存储），将 Object（可以理解为文件）存储到 bucket（可以理解为文件夹）里，你可以用 Swift 创建 container，然后上传文件，例如视频、照片，这些文件会被 replication 到不同服务器上以保证可靠性，Swift 可以不依靠虚拟机工作。所谓的云存储，OpenStack 就是用 Swift 实现的，类似于 Amazon AWS S3（Simple Storage Service）。

（2）Cinder 是 Block Storage（块存储），你可以把 Cinder 作为优盘管理程序来理解。你可以用 Cinder 创建 Volume，然后将它接到（attach）虚拟机上去，这个 Volume 就像虚拟机的一个存储分区一样工作。如果你把这个虚拟机 terminate 了，这个 Volume 和里面的数据依然还在，你还可以把它接到其他虚拟机上继续使用里面的数据。Cinder 创建的 Volume 必须被接到虚拟机上才能工作。类似于 Amazon AWS EBS（Elastic Block Storage）。

课后习题

1．OpenStack 的对象存储服务组件是（　　）。
A．MySQL　　　　B．Swift　　　　C．Cinder　　　　D．Glance
2．对象存储服务使用的数据库为（　　）。
A．SQL Server　　B．MySQL　　　C．Oracle　　　　D．SQLite
3．对于对象存储服务，（　　）不适合用它存储。
A．在线文档　　　B．镜像　　　　C．存档备份　　　D．图片
4．Swift 并不是文件系统或者实时的数据存储系统，它是_____。
5．Swift 使用的储存设备文件系统格式为_____。
6．怎样查看 Swift 服务储存状态？
7．如何校验 container.builder？

第 9 章

Web 服务 Dashboard

9.1　Dashboard 基本概念

Dashboard（Horizon）是一个 Web 接口，可使云平台管理员以及用户管理不同的 OpenStack 资源以及服务。Dashboard 提供了一个模块化的、基于 Web 的图形化界面服务门户。用户可以通过浏览器使用这个 Web 图形化界面来访问、控制其计算、存储和网络资源，如启动云主机、分配 IP 地址、设置访问控制等。

9.2　安装并配置 Dashboard

在控制节点上进行安装配置。

1. 安装 Dashboard 组件所需软件包

```
# yum install openstack-dashboard -y
```

2. 配置 Dashboard 组件

编辑 /etc/openstack-dashboard/local_settings 文件。
对以下行进行修改。
配置控制节点使用 Dashboard：

OPENSTACK_HOST = "controller"

配置允许所有主机访问 Dashboard：

ALLOWED_HOSTS = ['*',]

配置 memcached 的会话存储服务：

SESSION_ENGINE = 'django.contrib.sessions.backends.cache'

CACHES = {
　　'default': {
　　　　'BACKEND': 'django.core.cache.backends.memcached.MemcachedCache',
　　　　'LOCATION': 'controller:11211',
　　}
}

启用身份验证：

OPENSTACK_KEYSTONE_URL = "http://%s:5000/v3" % OPENSTACK_HOST

启用域的支持：

OPENSTACK_KEYSTONE_MULTIDOMAIN_SUPPORT = True

配置 API 版本：

OPENSTACK_API_VERSIONS = {
　　"identity": 3,
　　"image": 2,
　　"volume": 2,
}

配置域：

OPENSTACK_KEYSTONE_DEFAULT_DOMAIN = "default"

配置用户：

OPENSTACK_KEYSTONE_DEFAULT_ROLE = "user"

3. 启动 Dashboard 服务并设置开机自启动

systemctl restart httpd.service memcached.service

9.3 验证 Dashboard 服务

在浏览器地址栏中输入：192.168.100.10/dashboard。
输入域：default。
用户名：admin。
密码：******（自定义的 admin 用户的密码）。
即可登录 Dashboard，如图 9-1 所示。
成功登录界面如图 9-2 所示。
基本概况如图 9-3 所示。

图 9-1 登录 Dashboard

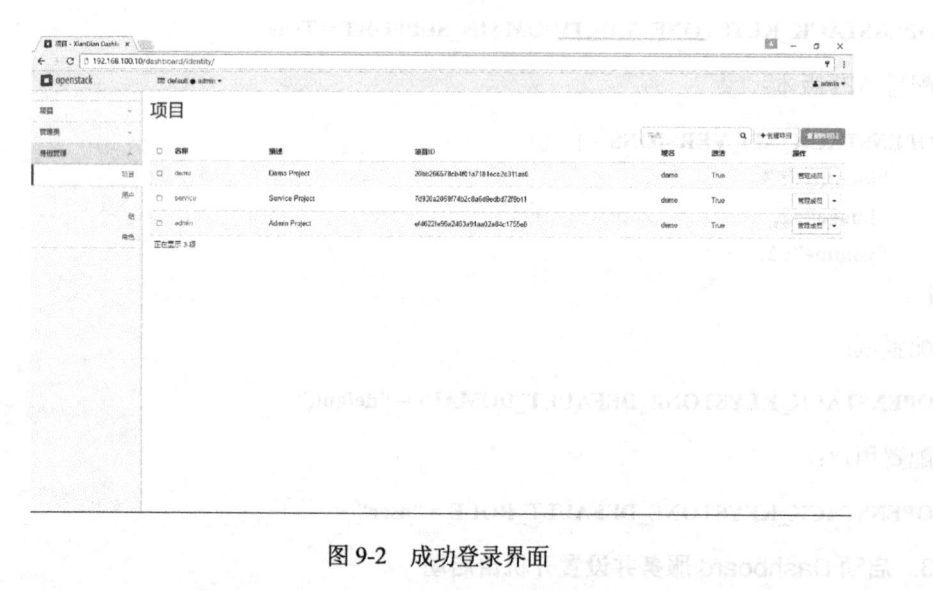

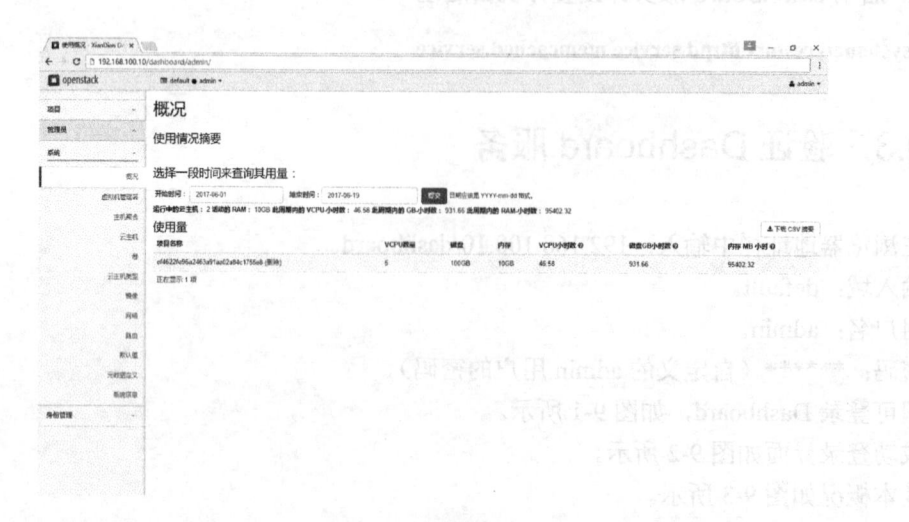

图 9-2 成功登录界面

图 9-3 基本概况

课后习题

1. Dashboard（Horizon）是一个_____接口，可使云平台管理员以及用户管理不同的_____资源以及服务。

2. Horizon 提供了一个模块化的、基于 Web 的图形化界面服务门户。用户可以通过浏览器使用这个 Web 图形化界面来访问、控制其_____、_____和_____。

第10章 块存储服务 Cinder

10.1 Cinder 基本概念

Cinder 从 OpenStack 的 Folsom 版本(于 2012 年 9 月发布)开始出现,用以替代 Nova-volume 服务,Cinder 为 OpenStack 提供了管理卷(Volume)的基础设施。

按 OpenStack 官方文档的表述,Cinder 是受请求得到、自助化访问的块储存服务,即 Cinder 有两个显著的特点:第一,必须由用户提出请求,才能得到该服务;第二,用户可以自定义的半自动化服务。Cinder 实现 LVM(逻辑卷管理),用以呈现存储资源给能够被 Nova 调用的端用户。简而言之,Cinder 虚拟化块存储设备池,提供端用户自助服务的 API 用以请求和使用这些块资源,并且不用了解存储的位置或设备信息。

OpenStack 块存储服务(Cinder)为虚拟机添加持久的存储,块存储提供一个基础设施为了管理卷,以及与 OpenStack 计算服务交互,为云主机提供卷。该服务也会激活管理卷的快照和卷类型的功能。

Cinder 架构如图 10-1 所示。

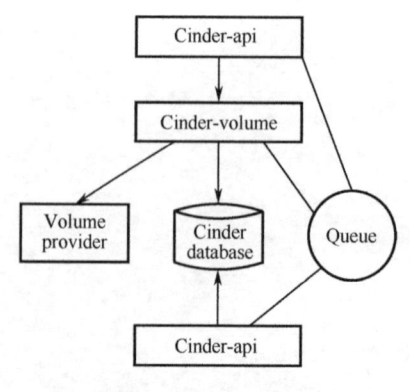

图 10-1 Cinder 架构

块存储服务通常包含下列组件。

Cinder-api：接收 API 请求，然后将之路由到 Cinder-volume 执行。

Cinder-volume：直接与块存储的服务交互，以及诸如 Cinder-scheduler 这样的流程交互，它和这些流程交互是通过消息队列实现的。Cinder-volume 服务响应那些发送到块存储服务的读/写请求以维护状态，它可以与多个存储供应商通过 Driver 架构做交互。

Cinder-scheduler：守护进程选择最佳的存储节点来创建卷，和它类似的组件是 Nova-scheduler。

消息队列：在块存储的进程之间路由信息。

Cinder iSCSI 实现原理详解如下。

iSCSI：Internet 小型计算机系统接口（iSCSI）是一种基于 TCP/IP 的协议，用来建立和管理 IP 存储设备、主机和客户机等之间的相互连接，并创建存储区域网络（SAN）。SAN 使得 SCSI 协议应用于高速数据传输网络成为可能，这种传输以数据块级别（block-level）在多个数据存储网络间进行。

10.2 安装并配置控制节点

10.2.1 数据库配置

登录 MySQL 并创建 Cinder 数据库：

mysql -uroot -p000000

创建 Cinder 数据库：

MariaDB [(none)]> **CREATE DATABASE cinder;**

设置授权用户和密码：

MariaDB [(none)]> **GRANT ALL PRIVILEGES ON cinder.* TO 'cinder'@'localhost' IDENTIFIED BY '000000';**

MariaDB [(none)]>**GRANT ALL PRIVILEGES ON cinder.* TO 'cinder'@'%' IDENTIFIED BY '000000';**

MariaDB [(none)]> **exit**

10.2.2 创建服务凭证和 API 端点

1. 生效 admin 用户环境变量

. admin-openrc

2. 创建服务凭证

创建名为 cinder 的 user：

openstack user create --domain default --password-prompt cinder

进行关联：

```
# openstack role add --project service --user cinder admin
```

创建 Cinder 服务实体认证 volume 和 volumev2：

```
# openstack service create --name cinder --description "OpenStack Block Storage" volume
# openstack service create --name cinderv2 --description "OpenStack Block Storage" volumev2
```

注：

```
[root@controller ~]# . admin-openrc
[root@controller ~]#
[root@controller ~]# openstack user create --domain default --password-prompt cinder
User Password:
Repeat User Password:
+-----------+----------------------------------+
| Field     | Value                            |
+-----------+----------------------------------+
| domain_id | d63fbba811b94cffb2cf9f88b59f4066 |
| enabled   | True                             |
| id        | 0de3a2839c64453a867cee71325984c7 |
| name      | cinder                           |
+-----------+----------------------------------+
[root@controller ~]# openstack role add --project service --user cinder admin
[root@controller ~]# openstack service create --name cinder \
>   --description "OpenStack Block Storage" volume
+-------------+----------------------------------+
| Field       | Value                            |
+-------------+----------------------------------+
| description | OpenStack Block Storage          |
| enabled     | True                             |
| id          | 8f0a88c04d4d45dda645bee2c94b0d29 |
| name        | cinder                           |
| type        | volume                           |
+-------------+----------------------------------+
[root@controller ~]# openstack service create --name cinderv2 \
>   --description "OpenStack Block Storage" volumev2
+-------------+----------------------------------+
| Field       | Value                            |
+-------------+----------------------------------+
| description | OpenStack Block Storage          |
| enabled     | True                             |
| id          | 5893b4bc5a7b46c3bbc4272c98da4e9e |
| name        | cinderv2                         |
| type        | volumev2                         |
+-------------+----------------------------------+
```

Cinder 服务创建了两个 Service，分别是 volume 和 volumev2。

3. 创建 API 端点

创建公共端点：

```
# openstack endpoint create --region RegionOne volume public http://controller: 8776/v1/%\(tenant_id\)s
# openstack endpoint create --region RegionOne volumev2 public http://controller: 8776/v2/%\(tenant_id\)s
```

创建外部端点

```
# openstack endpoint create --region RegionOne volume internal http://controller: 8776/v1/%\(tenant_id\)s
# openstack endpoint create --region RegionOne volumev2 internal http://controller: 8776/v2/%\(tenant_id\)s
```

创建管理端点

```
# openstack endpoint create --region RegionOne volume admin http://controller:8776/v1/%\(tenant_id\)s
# openstack endpoint create --region RegionOne volumev2 admin http://controller:8776/v2/%\(tenant_id\)s
```

```
[root@controller ~]# openstack endpoint create --region RegionOne \
>   volume public http://controller:8776/v1/%\(tenant_id\)s
+--------------+----------------------------------+
| Field        | Value                            |
+--------------+----------------------------------+
| enabled      | True                             |
| id           | 656b97bcf9a341e78fba6418eee747f7 |
| interface    | public                           |
| region       | RegionOne                        |
| region_id    | RegionOne                        |
| service_id   | 8f0a88c04d4d45dda645bee2c94b0d29 |
| service_name | cinder                           |
| service_type | volume                           |
| url          | http://controller:8776/v1/%(tenant_id)s |
+--------------+----------------------------------+
```

```
+-------------+------------------------------------------+
[root@controller ~]# openstack endpoint create --region RegionOne \
>    volume internal http://controller:8776/v1/%\(tenant_id\)s
+-------------+------------------------------------------+
| Field       | value                                    |
+-------------+------------------------------------------+
| enabled     | True                                     |
| id          | 4c32660bbba34cc1a45f16346a2a1a39         |
| interface   | internal                                 |
| region      | RegionOne                                |
| region_id   | RegionOne                                |
| service_id  | 8f0a88c04d4d45dda645bee2c94b0d29         |
| service_name| cinder                                   |
| service_type| volume                                   |
| url         | http://controller:8776/v1/%(tenant_id)s  |
+-------------+------------------------------------------+
[root@controller ~]#  openstack endpoint create --region RegionOne \
>    volume admin http://controller:8776/v1/%\(tenant_id\)s
+-------------+------------------------------------------+
| Field       | value                                    |
+-------------+------------------------------------------+
| enabled     | True                                     |
| id          | f82a79731d864b4a8372df542080e1ad         |
| interface   | admin                                    |
| region      | RegionOne                                |
| region_id   | RegionOne                                |
| service_id  | 8f0a88c04d4d45dda645bee2c94b0d29         |
| service_name| cinder                                   |
| service_type| volume                                   |
| url         | http://controller:8776/v1/%(tenant_id)s  |
+-------------+------------------------------------------+

[root@controller ~]# openstack endpoint create --region RegionOne \
>    volumev2 public http://controller:8776/v2/%\(tenant_id\)s
+-------------+------------------------------------------+
| Field       | value                                    |
+-------------+------------------------------------------+
| enabled     | True                                     |
| id          | cda08e69595d48699a1c8b0f0b76a3b1         |
| interface   | public                                   |
| region      | RegionOne                                |
| region_id   | RegionOne                                |
| service_id  | 5893b4bc5a7b46c3bbc4272c98da4e9e         |
| service_name| cinderv2                                 |
| service_type| volumev2                                 |
| url         | http://controller:8776/v2/%(tenant_id)s  |
+-------------+------------------------------------------+
[root@controller ~]# openstack endpoint create --region RegionOne \
>    volumev2 internal http://controller:8776/v2/%\(tenant_id\)s
+-------------+------------------------------------------+
| Field       | value                                    |
+-------------+------------------------------------------+
| enabled     | True                                     |
| id          | 68b62f66768f4dd7b1cb1f039baf7c22         |
| interface   | internal                                 |
| region      | RegionOne                                |
| region_id   | RegionOne                                |
| service_id  | 5893b4bc5a7b46c3bbc4272c98da4e9e         |
| service_name| cinderv2                                 |
| service_type| volumev2                                 |
| url         | http://controller:8776/v2/%(tenant_id)s  |
+-------------+------------------------------------------+
[root@controller ~]# openstack endpoint create --region RegionOne \
>    volumev2 admin http://controller:8776/v2/%\(tenant_id\)s
+-------------+------------------------------------------+
| Field       | value                                    |
+-------------+------------------------------------------+
| enabled     | True                                     |
| id          | 5941fcc2923c4b20886f18c13ed7d6ed         |
| interface   | admin                                    |
| region      | RegionOne                                |
| region_id   | RegionOne                                |
| service_id  | 5893b4bc5a7b46c3bbc4272c98da4e9e         |
| service_name| cinderv2                                 |
| service_type| volumev2                                 |
| url         | http://controller:8776/v2/%(tenant_id)s  |
+-------------+------------------------------------------+
```

10.2.3 安装并配置 Cinder 组件

1. 安装 Cinder 组件所需软件包

```
# yum install openstack-cinder -y
```

2. 配置 Cinder 所需组件

编辑/etc/cinder/cinder.conf 文件。

编辑[database]部分，配置数据库链接。

```
[database]
connection = mysql+pymysql://cinder:000000@controller/cinder
```

编辑[DEFAULT]和[oslo_messaging_rabbit]部分，配置 RabbitMQ 消息服务器链接。

```
[DEFAULT]
rpc_backend = rabbit

[oslo_messaging_rabbit]
rabbit_host = controller
rabbit_userid = openstack
rabbit_password = 000000
```

编辑[DEFAULT]和[keystone_authtoken]部分，配置 Keystone 身份认证。

```
[DEFAULT]
auth_strategy = keystone

[keystone_authtoken]
auth_uri = http://controller:5000
auth_url = http://controller:35357
memcached_servers = controller:11211
auth_type = password
project_domain_name = default
user_domain_name = default
project_name = service
username = cinder
password = 000000
```

编辑[DEFAULT]部分，配置控制节点管理 IP 地址。

```
[DEFAULT]
my_ip = 192.168.100.10
```

编辑[oslo_concurrency]部分，配置 loca_path。

```
[oslo_concurrency]
lock_path = /var/lib/cinder/tmp
```

3. 同步数据库

su -s /bin/sh -c "cinder-manage db sync" cinder

```
[root@controller ~]# su -s /bin/sh -c "cinder-manage db sync" cinder
Option "logdir" from group "DEFAULT" is deprecated. Use option "log-dir" from group "DEFAUL
2016-08-24 23:15:59.746 41255 WARNING py.warnings [-] /usr/lib/python2.7/site-packages/osl
ted
  exception.NotSupportedWarning

2016-08-24 23:16:01.202 41255 INFO migrate.versioning.api [-] 0 -> 1...
2016-08-24 23:16:01.757 41255 INFO migrate.versioning.api [-] done
2016-08-24 23:16:01.758 41255 INFO migrate.versioning.api [-] 1 -> 2...
2016-08-24 23:16:02.508 41255 INFO migrate.versioning.api [-] done
2016-08-24 23:16:02.508 41255 INFO migrate.versioning.api [-] 2 -> 3...
2016-08-24 23:16:02.571 41255 INFO migrate.versioning.api [-] done
2016-08-24 23:16:02.571 41255 INFO migrate.versioning.api [-] 3 -> 4...
2016-08-24 23:16:02.792 41255 INFO migrate.versioning.api [-] done
2016-08-24 23:16:02.792 41255 INFO migrate.versioning.api [-] 4 -> 5...
2016-08-24 23:16:02.834 41255 INFO migrate.versioning.api [-] done
2016-08-24 23:16:02.834 41255 INFO migrate.versioning.api [-] 5 -> 6...
2016-08-24 23:16:02.864 41255 INFO migrate.versioning.api [-] done
2016-08-24 23:16:02.865 41255 INFO migrate.versioning.api [-] 6 -> 7...
2016-08-24 23:16:02.912 41255 INFO migrate.versioning.api [-] done
2016-08-24 23:16:02.912 41255 INFO migrate.versioning.api [-] 7 -> 8...
2016-08-24 23:16:02.928 41255 INFO migrate.versioning.api [-] done
2016-08-24 23:16:02.928 41255 INFO migrate.versioning.api [-] 8 -> 9...
2016-08-24 23:16:02.954 41255 INFO migrate.versioning.api [-] done
2016-08-24 23:16:02.954 41255 INFO migrate.versioning.api [-] 9 -> 10...
2016-08-24 23:16:02.976 41255 INFO migrate.versioning.api [-] done
2016-08-24 23:16:02.976 41255 INFO migrate.versioning.api [-] 10 -> 11...
2016-08-24 23:16:03.021 41255 INFO migrate.versioning.api [-] done
2016-08-24 23:16:03.021 41255 INFO migrate.versioning.api [-] 11 -> 12...
2016-08-24 23:16:03.060 41255 INFO migrate.versioning.api [-] done
2016-08-24 23:16:03.060 41255 INFO migrate.versioning.api [-] 12 -> 13...
2016-08-24 23:16:03.085 41255 INFO migrate.versioning.api [-] done
2016-08-24 23:16:03.085 41255 INFO migrate.versioning.api [-] 13 -> 14...
2016-08-24 23:16:03.121 41255 INFO migrate.versioning.api [-] done
2016-08-24 23:16:03.122 41255 INFO migrate.versioning.api [-] 14 -> 15...
2016-08-24 23:16:03.137 41255 INFO migrate.versioning.api [-] done
2016-08-24 23:16:03.138 41255 INFO migrate.versioning.api [-] 15 -> 16...
2016-08-24 23:16:03.166 41255 INFO migrate.versioning.api [-] done
2016-08-24 23:16:03.166 41255 INFO migrate.versioning.api [-] 16 -> 17...
2016-08-24 23:16:03.267 41255 INFO migrate.versioning.api [-] done
2016-08-24 23:16:03.267 41255 INFO migrate.versioning.api [-] 17 -> 18...
2016-08-24 23:16:03.323 41255 INFO migrate.versioning.api [-] done
2016-08-24 23:16:03.323 41255 INFO migrate.versioning.api [-] 18 -> 19...
2016-08-24 23:16:03.362 41255 INFO migrate.versioning.api [-] done
2016-08-24 23:16:03.362 41255 INFO migrate.versioning.api [-] 19 -> 20...
2016-08-24 23:16:03.383 41255 INFO migrate.versioning.api [-] done
2016-08-24 23:16:03.384 41255 INFO migrate.versioning.api [-] 20 -> 21...
2016-08-24 23:16:03.406 41255 INFO migrate.versioning.api [-] done
2016-08-24 23:16:03.406 41255 INFO migrate.versioning.api [-] 21 -> 22...
2016-08-24 23:16:03.437 41255 INFO migrate.versioning.api [-] done
2016-08-24 23:16:03.437 41255 INFO migrate.versioning.api [-] 22 -> 23...
2016-08-24 23:16:03.462 41255 INFO migrate.versioning.api [-] done
2016-08-24 23:16:03.462 41255 INFO migrate.versioning.api [-] 23 -> 24...
```

注：进入 Cinder 数据库查看是否有数据表，验证是否同步成功。

4. 配置 Nova 服务使用 Cinder

编辑 **/etc/nova/nova.conf** 文件。

编辑 **[cinder]** 部分，配置使用。

```
[cinder]
os_region_name = RegionOne
```

5. 启动 Cinder 服务并设置开机自启动

systemctl restart openstack-nova-api.service
systemctl enable openstack-cinder-api.service openstack-cinder-scheduler.service
systemctl start openstack-cinder-api.service openstack-cinder-scheduler.service

10.3 安装并配置存储节点

注：为简单起见，这里配置一个存储节点，包含两个空本地块存储设备/dev/sdb 和/dev/sdc。
注：因为又添加了一个存储节点，所以需要按照本书的第 3 章、第 4 章、第 6 章、第 7 章做准备环境的操作。
（1）存储节点 1 的管理 IP 地址为 192.168.100.30；主机名为 storage1。
（2）配置 hosts 文件配对，对所有的主机都需要进行更新配置。
（3）安全配置（关闭防火墙、Selinux）、配置 yum 源、NTP 以及安装 OpenStack 包。

10.3.1 安装工具包

1. 安装并启动

`# yum install lvm2 -y`

启动 lvm2 并设置开机自启动。

`# systemctl enable lvm2-lvmetad.service`
`# systemctl start lvm2-lvmetad.service`

2. 创建物理卷/dev/sdb

`# pvcreate /dev/sdb`

```
[root@compute ~]#  pvcreate /dev/sdb
  Physical volume "/dev/sdb" successfully created
```

3. 创建卷组 cinder-volumes

`# vgcreate cinder-volumes /dev/sdb`

```
[root@compute ~]# vgcreate cinder-volumes /dev/sdb
  Volume group "cinder-volumes" successfully created
```

注：创建之前检查是否挂载有空硬盘。

`# fdisk -l`

```
Disk /dev/sdb: 10.7 GB, 10737418240 bytes, 20971520 sectors
Units = sectors of 1 * 512 = 512 bytes
Sector size (logical/physical): 512 bytes / 512 bytes
I/O size (minimum/optimal): 512 bytes / 512 bytes
```

4. 配置 lvm2 组件

编辑**/etc/lvm/lvm.conf** 文件，配置过滤器。
编辑**# Configuration section devices** 部分。
添加：

`filter = [ "a/sdb/", "r/.*/"]`

注：以实际硬盘名称为准。

10.3.2 安装并配置组件

1. 安装 Cinder 组件所需软件包

yum install openstack-cinder targetcli python-keystone -y

2. 配置 Cinder 所需组件

编辑**/etc/cinder/cinder.conf** 文件。

编辑**[database]**部分，配置数据库链接。

[database]
connection = mysql+pymysql://cinder:000000@controller/cinder

编辑**[DEFAULT]**和**[oslo_messaging_rabbit]**部分，配置 RabbitMQ 消息服务器链接。

[DEFAULT]
rpc_backend = rabbit

[oslo_messaging_rabbit]
rabbit_host = controller
rabbit_userid = openstack
rabbit_password = 000000

编辑**[DEFAULT]**和**[keystone_authtoken]**部分，配置 Keystone 身份认证。

[DEFAULT]
auth_strategy = keystone

[keystone_authtoken]

auth_uri = http://controller:5000
auth_url = http://controller:35357
memcached_servers = controller:11211
auth_type = password
project_domain_name = default
user_domain_name = default
project_name = service
username = cinder
password = 000000

编辑[DEFAULT]部分，配置存储节点管理 IP 地址。

[DEFAULT]
my_ip = 192.168.100.30

编辑**[lvm]**部分，配置 lvm 后端，以及基于 TCP/IP 的协议的（iSCSI）接口和相对应的服务。

[lvm]
volume_driver = cinder.volume.drivers.lvm.LVMVolumeDriver
volume_group = cinder-volumes

```
iscsi_protocol = iscsi
iscsi_helper = lioadm
```

注：如果在配置文件中没找到[lvm]部分，则需要自己添加。

编辑**[DEFAULT]**部分，启用 LVM 后端。

```
[DEFAULT]
enabled_backends = lvm
```

编辑**[DEFAULT]**部分，配置 Glance 服务 API。

```
[DEFAULT]
glance_api_servers = http://controller:9292
```

编辑**[oslo_concurrency]**部分，配置 lock_path。

```
[oslo_concurrency]
lock_path = /var/lib/cinder/tmp
```

3. 启动 Cinder 服务并设置开机自启动

```
# systemctl enable openstack-cinder-volume.service target.service
# systemctl start openstack-cinder-volume.service target.service
```

10.4 验证 Cinder 服务

控制节点验证：

1. 生效 admin 用户环境变量

```
# . admin-openrc
```

2. 查看 Cinder 服务

```
# cinder service-list
+------------------+--------------+------+---------+-------+----------------------------+-----------------+
|     Binary       |     Host     | Zone | Status  | State |         Updated_at         | Disabled Reason |
+------------------+--------------+------+---------+-------+----------------------------+-----------------+
| cinder-scheduler |  controller  | nova | enabled |  up   | 2017-06-19T03:29:00.000000 |        -        |
|  cinder-volume   | compute@lvm  | nova | enabled |  up   | 2017-06-19T03:29:03.000000 |        -        |
+------------------+--------------+------+---------+-------+----------------------------+-----------------+
```

课后习题

1. 在 Cinder 架构中，负责接收 API 请求的服务是（　　）。
   A. Cinder-api　　　　　　　　　　B. Cinder-volume
   C. Cinder-scheduler　　　　　　　D. Nova-scheduler

2. Cinder 组件的配置文件为（　　）。
   A. /etc/cinder　　　　　　　　　　B. /etc/lvm
   C. /etc/cinder/cinder.conf　　　　D. /etc/lvm/lvm.conf

3. Cinder 组件含有的服务个数为（　　）。
A．3　　　　　　B．4　　　　　　　C．5　　　　　　D．6
4. 在安装 Cinder 过程中，同步数据库的命令为_____。
5. 查看 Cinder 服务的命令为_____。
6. 查看卷列表的命令为_____。
7. 如何创建 Cinder 服务的公共端点？

第11章

编配服务 Heat

11.1 Heat 基本概念

Heat 是一个基于模板来编排复合云应用的服务。它目前支持亚马逊的 CloudFormation 模板格式，也支持 Heat 自有的 Hot 模板格式。模板的使用简化了复杂基础设施，服务和应用的定义和部署。模板支持丰富的资源类型，不仅覆盖了常用的基础架构，包括计算、网络、存储、镜像，还覆盖了像 Ceilometer 的警报、Sahara 的集群、Trove 的云主机等高级资源。

Heat 服务包含以下重要的组件。

Heat-api 组件：实现 OpenStack 天然支持的 REST API。该组件通过把 API 请求经由 AMQP 传送给 Heat-Engine 来处理 API 请求。

Heat-api-cfn 组件：提供兼容 AWS CloudFormation 的 API，同时也会把 API 请求通过 AMQP 转发给 Heat-Engine。

Heat-Engine 组件：提供 Heat 最主要的协作功能。

Heat 架构图：如图 11-1 所示。

说明：

用户在 Horizon 中或者命令行中提交包含模板和参数输入的请求，Horizon 或者命令行工具会把请求转化为 REST 格式的 API 调用，然后调用 Heat-api 或者 Heat-api-cfn。Heat-api 和 Heat-api-cfn 会验证模板的正确性，然后通过 AMQP 异步传递给 Heat Engine 来处理请求。

当 Heat Engine 收到请求后，会把请求解析为各种类型的资源，每种资源都对应 OpenStack 其他的服务客户端，然后通过发送 REST 的请求给其他服务。通过如此的解析和协作，最终完成请求的处理。

Heat Engine 在这里的作用分为三层：第一层处理 Heat 层面的请求，即根据模板和输入参数来创建 Stack，这里的 Stack 由各种资源组合而成。第二层解析 Stack 里各种资源的依赖关系，

Stack 和嵌套 Stack 的关系。第三层根据解析出来的关系，依次调用各种服务客户端来创建各种资源。

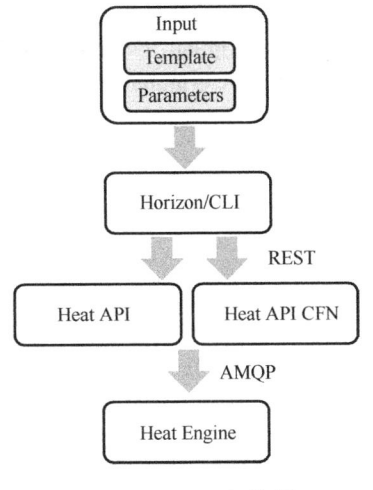

图 11-1 Heat 架构图

11.2 数据库配置

登录 MySQL 并创建 Glance 数据库：

mysql -uroot -p000000

创建 Glance 数据库：

MariaDB [(none)]>CREATE DATABASE heat;

设置授权用户和密码：

MariaDB [(none)]> GRANT ALL PRIVILEGES ON heat.* TO 'heat'@'localhost' IDENTIFIED BY '000000';
MariaDB [(none)]>GRANT ALL PRIVILEGES ON heat.* TO 'heat'@'%' IDENTIFIED BY '000000';
MariaDB [(none)]>exit

11.3 创建服务凭证和 API 端点

1. 生效 admin 用户环境变量

. admin-openrc

2. 创建服务凭证

创建名为 heat 的 user：

openstack user create --domain default --password-prompt heat

进行关联：

openstack role add --project service --user heat admin

创建 Heat 服务实体认证 heat 和 heat-cfn：

openstack service create --name heat --description "Orchestration" orchestration
openstack service create --name heat-cfn --description "Orchestration" cloudformation

```
[root@controller ~]# openstack role add --project service --user heat admin
[root@controller ~]# openstack service create --name heat \
>  --description "Orchestration" orchestration
+-------------+----------------------------------+
| Field       | Value                            |
+-------------+----------------------------------+
| description | Orchestration                    |
| enabled     | True                             |
| id          | 10e68d7dc2fa40c282c9db8e8d2bda94 |
| name        | heat                             |
| type        | orchestration                    |
+-------------+----------------------------------+
[root@controller ~]# openstack service create --name heat-cfn \
>  --description "Orchestration" cloudformation
+-------------+----------------------------------+
| Field       | Value                            |
+-------------+----------------------------------+
| description | Orchestration                    |
| enabled     | True                             |
| id          | b02dca65db534a13aac2b4edb987cf06 |
| name        | heat-cfn                         |
| type        | cloudformation                   |
+-------------+----------------------------------+
```

3. 创建 API 端点

创建公共端点：

openstack endpoint create --region RegionOne orchestration public http://controller:8004/v1/%\(tenant_id\)s
openstack endpoint create --region RegionOne cloudformation public http://controller:8000/v1

创建外部端点：

openstack endpoint create --region RegionOne orchestration internal http://controller:8004/v1/%\(tenant_id\)s
openstack endpoint create --region RegionOne cloudformation internal http://controller:8000/v1

创建管理端点：

openstack endpoint create --region RegionOne orchestration admin http://controller:8004/v1/%\(tenant_id\)s
openstack endpoint create --region RegionOne cloudformation admin http://controller:8000/v1

```
[root@controller ~]# openstack endpoint create --region RegionOne \
>  orchestration public http://controller:8004/v1/%\(tenant_id\)s
+--------------+----------------------------------------------+
| Field        | Value                                        |
+--------------+----------------------------------------------+
| enabled      | True                                         |
| id           | 9db39fb0f2ee4327a64da1b5fc6b7970             |
| interface    | public                                       |
| region       | RegionOne                                    |
| region_id    | RegionOne                                    |
| service_id   | 10e68d7dc2fa40c282c9db8e8d2bda94             |
| service_name | heat                                         |
| service_type | orchestration                                |
| url          | http://controller:8004/v1/%(tenant_id)s      |
+--------------+----------------------------------------------+
[root@controller ~]# openstack endpoint create --region RegionOne \
>  orchestration internal http://controller:8004/v1/%\(tenant_id\)s
+--------------+----------------------------------------------+
| Field        | Value                                        |
+--------------+----------------------------------------------+
| enabled      | True                                         |
| id           | e252573ce7a24871b642e1cedcd34cce             |
| interface    | internal                                     |
| region       | RegionOne                                    |
| region_id    | RegionOne                                    |
| service_id   | 10e68d7dc2fa40c282c9db8e8d2bda94             |
| service_name | heat                                         |
| service_type | orchestration                                |
| url          | http://controller:8004/v1/%(tenant_id)s      |
+--------------+----------------------------------------------+
```

```
[root@controller ~]# openstack endpoint create --region RegionOne \
> orchestration admin http://controller:8004/v1/%\(tenant_id\)s
+--------------+------------------------------------------+
| Field        | value                                    |
+--------------+------------------------------------------+
| enabled      | True                                     |
| id           | 0e03238de546453a8e0e6d29ce4c2095         |
| interface    | admin                                    |
| region       | RegionOne                                |
| region_id    | RegionOne                                |
| service_id   | 10e68d7dc2fa40c282c9db8e8d2bda94         |
| service_name | heat                                     |
| service_type | orchestration                            |
| url          | http://controller:8004/v1/%(tenant_id)s  |
+--------------+------------------------------------------+
[root@controller ~]# openstack endpoint create --region RegionOne \
> cloudformation public http://controller:8000/v1
+--------------+------------------------------------------+
| Field        | value                                    |
+--------------+------------------------------------------+
| enabled      | True                                     |
| id           | 4b3798ed313649459fd34eb129bf684b         |
| interface    | public                                   |
| region       | RegionOne                                |
| region_id    | RegionOne                                |
| service_id   | b02dca65db534a13aac2b4edb987cf06         |
| service_name | heat-cfn                                 |
| service_type | cloudformation                           |
| url          | http://controller:8000/v1                |
+--------------+------------------------------------------+
[root@controller ~]# openstack endpoint create --region RegionOne \
> cloudformation internal http://controller:8000/v1
+--------------+------------------------------------------+
| Field        | value                                    |
+--------------+------------------------------------------+
| enabled      | True                                     |
| id           | 7520b1ac21f74ec0afcb33b7ae13982b         |
| interface    | internal                                 |
| region       | RegionOne                                |
| region_id    | RegionOne                                |
| service_id   | b02dca65db534a13aac2b4edb987cf06         |
| service_name | heat-cfn                                 |
| service_type | cloudformation                           |
| url          | http://controller:8000/v1                |
+--------------+------------------------------------------+
[root@controller ~]# openstack endpoint create --region RegionOne \
> cloudformation admin http://controller:8000/v1
+--------------+------------------------------------------+
| Field        | value                                    |
+--------------+------------------------------------------+
| enabled      | True                                     |
| id           | 5980fd5813c349c299b702dbc38d4389         |
| interface    | admin                                    |
| region       | RegionOne                                |
| region_id    | RegionOne                                |
| service_id   | b02dca65db534a13aac2b4edb987cf06         |
| service_name | heat-cfn                                 |
| service_type | cloudformation                           |
| url          | http://controller:8000/v1                |
+--------------+------------------------------------------+
```

4. 配置 Heat 身份管理服务

创建 Heat 服务的 domain：

 # openstack domain create --description "Stack projects and users" heat

创建名为 heat_domain_admin 的 user：

 # openstack user create --domain heat --password-prompt heat_domain_admin

进行关联：

 # openstack role add --domain heat --user-domain heat --user heat_domain_admin admin

```
[root@controller ~]# openstack role create heat_stack_owner
+-----------+----------------------------------+
| Field     | Value                            |
+-----------+----------------------------------+
| domain_id | None                             |
| id        | c207e76bbd8146e3849100ea41c603fe |
| name      | heat_stack_owner                 |
+-----------+----------------------------------+
[root@controller ~]# openstack role add --project demo --user demo heat_stack_owner
[root@controller ~]# openstack role create heat_stack_user
+-----------+----------------------------------+
| Field     | Value                            |
+-----------+----------------------------------+
| domain_id | None                             |
| id        | 3b0d71da4fda4c469ff333e7c168fb0a |
| name      | heat_stack_user                  |
+-----------+----------------------------------+
```

创建名为 heat_stack_owner 的 role：

openstack role create heat_stack_owner

进行关联：

openstack role add --project demo --user demo heat_stack_owner

创建名为 heat_stack_user 的 role：

openstack role create heat_stack_user

```
[root@controller ~]# openstack domain create --description "Stack projects and users" heat
+-------------+----------------------------------+
| Field       | Value                            |
+-------------+----------------------------------+
| description | Stack projects and users         |
| enabled     | True                             |
| id          | 37893facd9a5436284ca6414bab53da4 |
| name        | heat                             |
+-------------+----------------------------------+
[root@controller ~]# openstack user create --domain heat --password-prompt heat_domain_admin
User Password:
Repeat User Password:
+-----------+----------------------------------+
| Field     | Value                            |
+-----------+----------------------------------+
| domain_id | 37893facd9a5436284ca6414bab53da4 |
| enabled   | True                             |
| id        | 2a63ec647a6445f5a148b0cb4b863e45 |
| name      | heat_domain_admin                |
+-----------+----------------------------------+
[root@controller ~]# openstack role add --domain heat --user-domain heat --user heat_domain_admin admin
```

11.4 安装并配置 Heat

1. 安装 Heat 所需软件包

yum install openstack-heat-api openstack-heat-api-cfn openstack-heat-engine -y

2. 配置 Heat 所需组件

编辑 **/etc/heat/heat.conf** 文件。

编辑 **[database]** 部分，配置数据库链接。

[database]
connection = mysql+pymysql://heat:000000@controller/heat

编辑 **[DEFAULT]** 和 **[oslo_messaging_rabbit]** 部分，配置 RabbitMQ 消息服务器链接。

[DEFAULT]
rpc_backend = rabbit

[oslo_messaging_rabbit]
rabbit_host = controller
rabbit_userid = openstack
rabbit_password = 000000

编辑 **[keystone_authtoken]**、**[trustee]**、**[clients_keystone]**、**[ec2authtoken]** 部分，配置 Keystone 认证。

[keystone_authtoken]

```
auth_uri = http://controller:5000
auth_url = http://controller:35357
memcached_servers = controller:11211
auth_type = password
project_domain_name = default
user_domain_name = default
project_name = service
username = heat
password = 000000

[trustee]
auth_plugin = password
auth_url = http://controller:35357
username = heat
password = 000000
user_domain_name = default

[clients_keystone]
auth_uri = http://controller:35357

[ec2authtoken]
auth_uri = http://controller:5000
```

编辑[DEFAULT]部分，配置 metadata 和等待状态的网址。

```
[DEFAULT]
heat_metadata_server_url = http://controller:8000
heat_waitcondition_server_url = http://controller:8000/v1/waitcondition
```

编辑[DEFAULT]部分，配置域和管理凭据。

```
[DEFAULT]
stack_domain_admin = heat_domain_admin
stack_domain_admin_password = 000000
stack_user_domain_name = heat
```

3. 同步数据库

su -s /bin/sh -c "heat-manage db_sync" heat

```
[root@controller ~]# su -s /bin/sh -c "heat-manage db_sync" heat
2016-08-25 02:05:37.788 43883 INFO migrate.versioning.api [-] 27 -> 28...
2016-08-25 02:05:38.122 43883 INFO migrate.versioning.api [-] done
2016-08-25 02:05:38.122 43883 INFO migrate.versioning.api [-] 28 -> 29...
2016-08-25 02:05:38.230 43883 INFO migrate.versioning.api [-] done
2016-08-25 02:05:38.231 43883 INFO migrate.versioning.api [-] 29 -> 30...
2016-08-25 02:05:38.260 43883 INFO migrate.versioning.api [-] done
2016-08-25 02:05:38.260 43883 INFO migrate.versioning.api [-] 30 -> 31...
2016-08-25 02:05:38.287 43883 INFO migrate.versioning.api [-] done
```

注：进入 Heat 数据库查看是否有数据表，验证是否同步成功。

4. 启动 Heat 服务并设置开机自启动

systemctl enable openstack-heat-api.service openstack-heat-api-cfn.service openstack-heat-engine.service

systemctl start openstack-heat-api.service openstack-heat-api-cfn.service openstack-heat-engine.service

11.5 验证 Heat

1. 生效 admin 用户环境变量

`# . admin-openrc`

2. 查看服务

`# openstack orchestration service list`

```
[root@controller ~]# . admin-openrc
[root@controller ~]# openstack orchestration service list
+------------+-------------+--------------------------------------+------------+--------+----------------------------+--------+
| hostname   | binary      | engine_id                            | host       | topic  | updated_at                 | status |
+------------+-------------+--------------------------------------+------------+--------+----------------------------+--------+
| controller | heat-engine | f8afff9c-e431-4e5e-961f-dd1752df0363 | controller | engine | 2016-08-25T06:12:31.000000 | up     |
| controller | heat-engine | bf3f028d-e45a-4d2d-a031-21049f4b6a24 | controller | engine | 2016-08-25T06:12:31.000000 | up     |
| controller | heat-engine | ee9f3603-e041-484f-a987-fe832e7e9084 | controller | engine | 2016-08-25T06:12:31.000000 | up     |
| controller | heat-engine | 1a8e9a99-8abe-42e1-bde6-d1f8ef1efd77 | controller | engine | 2016-08-25T06:12:31.000000 | up     |
```

课后习题

1. 在下列服务中，不属于 Heat 组件的是（   ）。
   A．Heat-api   B．Heat-api-cfn   C．Heat-volume   D．Heat-engine
2. 在 Heat 组件中，公共端点的数量为（   ）。
   A．2   B．3   C．4   D．1
3. 启动 Heat 组件的相关服务命令为_____。
4. 如何查看 Heat 服务？

第12章

运行云主机

12.1 创建云网络

12.1.1 Provider Network

在启动云主机之前，必须创建一些必要的虚拟网络基础设施。一个云主机使用 provider（external）网络，通过 L2（桥接/交换机）连接到物理网络基础设施。这个网络包括 DHCP 服务即给云主机提供 IP 地址。

创建网络：

1. **生效 admin 用户环境变量**

```
# . admin-openrc
```

2. **创建网络**

```
# neutron net-create --shared --provider:physical_network provider --provider:network_type flat provider
```

```
[root@controller ~]# neutron net-create --shared --provider:physical_network provider \
> --provider:network_type flat provider
Created a new network:
+---------------------------+--------------------------------------+
| Field                     | value                                |
+---------------------------+--------------------------------------+
| admin_state_up            | True                                 |
| availability_zone_hints   |                                      |
| availability_zones        |                                      |
| created_at                | 2016-08-25T06:32:36                  |
| description               |                                      |
| id                        | 8cba188d-cf92-4a1f-bdf3-0cbfb3a6e06e |
| ipv4_address_scope        |                                      |
| ipv6_address_scope        |                                      |
| mtu                       | 1500                                 |
| name                      | provider                             |
| port_security_enabled     | True                                 |
| provider:network_type     | flat                                 |
| provider:physical_network | provider                             |
| provider:segmentation_id  |                                      |
| router:external           | False                                |
| shared                    | True                                 |
| status                    | ACTIVE                               |
| subnets                   |                                      |
| tags                      |                                      |
| tenant_id                 | e71744580698483c90b1ac0a03d66313     |
| updated_at                | 2016-08-25T06:32:37                  |
+---------------------------+--------------------------------------+
```

注：

--shared 选项：允许所有项目使用虚拟网络。

--provider:physical_network 选项和**--provider:network_type** 选项使虚拟网络和物理网络连接。

3. 创建子网

```
# neutron subnet-create --name provider \
>     --allocation-pool start=192.168.200.100,end=192.168.200.200\
>     --dns-nameserver 8.8.4.4 --gateway 192.168.200.1 \
> provider 192.168.200.0/24
```

```
[root@controller ~]# neutron subnet-create --name provider \
>     --allocation-pool start=192.168.200.100,end=192.168.200.200\
>     --dns-nameserver 8.8.4.4 --gateway 192.168.200.1 \
>   provider 192.168.200.0/24
Created a new subnet:
+-------------------+--------------------------------------------------+
| Field             | Value                                            |
+-------------------+--------------------------------------------------+
| allocation_pools  | {"start": "192.168.200.100", "end": "192.168.200.200"} |
| cidr              | 192.168.200.0/24                                 |
| created_at        | 2016-08-25T10:21:55                              |
| description       |                                                  |
| dns_nameservers   | 8.8.4.4                                          |
| enable_dhcp       | True                                             |
| gateway_ip        | 192.168.200.1                                    |
| host_routes       |                                                  |
| id                | e0e64a5d-e07a-4db5-87a5-7b59e2611d04             |
| ip_version        | 4                                                |
| ipv6_address_mode |                                                  |
| ipv6_ra_mode      |                                                  |
| name              | provider                                         |
| network_id        | 18a59b48-85d4-4b31-9e11-d466f5341c5b             |
| subnetpool_id     |                                                  |
| tenant_id         | e71744580698483c90b1ac0a03d66313                 |
| updated_at        | 2016-08-25T10:21:55                              |
+-------------------+--------------------------------------------------+
```

注：

--allocation-pool 选项：这是 IP 地址池。

--dns-nameserver 选项：DNS 服务。

--gateway 选项：网关。

192.168.200.0/24 表示网络号。

12.1.2　Self-service Network

创建 self-service network 之前必须创建 Provider network。

创建网络：

1. 生效 demo 用户环境变量

`# . demo-openrc`

2. 创建网络

`# neutron net-create selfservice`

```
[root@controller ~]# neutron net-create selfservice
Created a new network:
+-------------------------+--------------------------------------+
| Field                   | value                                |
+-------------------------+--------------------------------------+
| admin_state_up          | True                                 |
| availability_zone_hints |                                      |
| availability_zones      |                                      |
| created_at              | 2016-08-25T10:25:13                  |
| description             |                                      |
| id                      | 8f82ab7d-0cd5-4c79-adc4-e4365fa6ff11 |
| ipv4_address_scope      |                                      |
| ipv6_address_scope      |                                      |
| mtu                     | 1450                                 |
| name                    | selfservice                          |
| port_security_enabled   | True                                 |
| router:external         | False                                |
| shared                  | False                                |
| status                  | ACTIVE                               |
| subnets                 |                                      |
| tags                    |                                      |
| tenant_id               | 622d72c12b9b4fb49b76860b3178f490     |
| updated_at              | 2016-08-25T10:25:14                  |
+-------------------------+--------------------------------------+
```

3. 创建子网

neutron subnet-create --name selfservice \
> --dns-nameserver 8.8.4.4 --gateway 10.0.0.1 \
> selfservice 10.0.0.0/24

```
[root@controller ~]# neutron subnet-create --name selfservice \
>    --dns-nameserver 8.8.4.4 --gateway 10.0.0.1 \
>    selfservice 10.0.0.0/24
Created a new subnet:
+-------------------+------------------------------------------+
| Field             | value                                    |
+-------------------+------------------------------------------+
| allocation_pools  | {"start": "10.0.0.2", "end": "10.0.0.254"} |
| cidr              | 10.0.0.0/24                              |
| created_at        | 2016-08-25T10:26:55                      |
| description       |                                          |
| dns_nameservers   | 8.8.4.4                                  |
| enable_dhcp       | True                                     |
| gateway_ip        | 10.0.0.1                                 |
| host_routes       |                                          |
| id                | 040a3afc-6b8b-49f5-af09-72fe990924bd     |
| ip_version        | 4                                        |
| ipv6_address_mode |                                          |
| ipv6_ra_mode      |                                          |
| name              | selfservice                              |
| network_id        | 8f82ab7d-0cd5-4c79-adc4-e4365fa6ff11     |
| subnetpool_id     |                                          |
| tenant_id         | 622d72c12b9b4fb49b76860b3178f490         |
| updated_at        | 2016-08-25T10:26:55                      |
+-------------------+------------------------------------------+
```

注：此处网络创建没有设置地址池，默认为全部；网关和网络号是自定义的云主机网络，自由设置，但是不要和物理机相同。

4. 创建路由

生效 admin 用户环境变量：

. admin-openrc

设置 provider 网络为外部网络：

neutron net-update provider --router:external

```
[root@controller ~]#  neutron net-update provider --router:external
Updated network: provider
```

生效 demo 用户环境变量：

#. demo-openrc

创建路由器：

neutron router-create router

```
[root@controller ~]# neutron router-create router
Created a new router:
+-------------------------+--------------------------------------+
| Field                   | Value                                |
+-------------------------+--------------------------------------+
| admin_state_up          | True                                 |
| availability_zone_hints |                                      |
| availability_zones      |                                      |
| description             |                                      |
| distributed             | False                                |
| external_gateway_info   |                                      |
| ha                      | False                                |
| id                      | d0d1ee79-cf3a-42a4-95d2-c4caf3c4e62a |
| name                    | router                               |
| routes                  |                                      |
| status                  | ACTIVE                               |
| tenant_id               | e71744580698483c90b1ac0a03d66313     |
+-------------------------+--------------------------------------+
```

添加 self-service 网络的子网的路由器口：

neutron router-interface-add router selfservice

```
[root@controller ~]# neutron router-interface-add router selfservice
Added interface 301ce51f-866e-4610-a5af-83633e69f1c5 to router router.
```

设置路由器的 provider 网络的网关：

neutron router-gateway-set router provider

```
[root@controller ~]# neutron router-gateway-set router provider
Set gateway for router router
```

12.1.3 验证网络

生效 admin 用户环境变量：

#. admin-openrc

查看网络命名空间：

ip netns

```
[root@controller ~]# ip netns
qrouter-adea2a28-9bf6-4862-9cbe-9c3da392e244 (id: 2)
qdhcp-8f82ab7d-0cd5-4c79-adc4-e4365fa6ff11 (id: 1)
qdhcp-18a59b48-85d4-4b31-9e11-d466f5341c5b (id: 0)
```

列出路由器端口地址，查看 provider 网络网关 IP 地址

neutron router-port-list router

```
[root@controller ~]# neutron router-port-list router
+--------------------------------------+------+-------------------+------------------------------------------------------------------------------------+
| id                                   | name | mac_address       | fixed_ips                                                                          |
+--------------------------------------+------+-------------------+------------------------------------------------------------------------------------+
| 1efe2048-1f9e-4369-9260-6c05f19366c9 |      | fa:16:3e:80:70:c7 | {"subnet_id": "e0e64a5d-e07a-4db5-87a5-7b59e2611d04", "ip_address": "192.168.200.101"} |
| 301ce51f-866e-4610-a5af-83633e69f1c5 |      | fa:16:3e:c9:17:3f | {"subnet_id": "040a3afc-6b8b-49f5-af09-72fe990924bd", "ip_address": "10.0.0.1"}    |
+--------------------------------------+------+-------------------+------------------------------------------------------------------------------------+
```

12.2 创建云主机

12.2.1 设置密钥对

生效 demo 用户环境变量：

 # . demo-openrc

创建密钥对：

 # ssh-keygen -q -N ""

```
[root@controller ~]# ssh-keygen -q -N ""
Enter file in which to save the key (/root/.ssh/id_rsa):
```

此处直接回车。

 # openstack keypair create --public-key ~/.ssh/id_rsa.pub mykey

```
[root@controller ~]# openstack keypair create --public-key ~/.ssh/id_rsa.pub mykey
+-------------+-------------------------------------------------+
| Field       | value                                           |
+-------------+-------------------------------------------------+
| fingerprint | 36:58:99:a7:df:48:b5:40:dc:4b:ec:54:6a:34:c2:26 |
| name        | mykey                                           |
| user_id     | d5c9272aadc24c00899fe2fe00f3aa7f                |
+-------------+-------------------------------------------------+
```

查看密钥对：

 # openstack keypair list

```
[root@controller ~]# openstack keypair list
+-------+-------------------------------------------------+
| Name  | Fingerprint                                     |
+-------+-------------------------------------------------+
| mykey | 36:58:99:a7:df:48:b5:40:dc:4b:ec:54:6a:34:c2:26 |
+-------+-------------------------------------------------+
```

12.2.2 添加安全规则

为默认安全组 default 添加规则。

允许 ICMP（ping）：

 # openstack security group rule create --proto icmp default

```
[root@controller ~]# openstack security group rule create --proto icmp default
+-------------------+--------------------------------------+
| Field             | value                                |
+-------------------+--------------------------------------+
| id                | eaec29ba-f3f2-4b3e-80ff-0a77792eeb74 |
| ip_protocol       | icmp                                 |
| ip_range          | 0.0.0.0/0                            |
| parent_group_id   | 96dd7907-f70e-4973-bf79-623fb6fe4e1e |
| port_range        |                                      |
| remote_security_group |                                  |
+-------------------+--------------------------------------+
```

允许 SSH 访问：

 # openstack security group rule create --proto tcp --dst-port 22 default

```
[root@controller ~]# openstack security group rule create --proto tcp --dst-port 22 default
+-------------------+--------------------------------------+
| Field             | Value                                |
+-------------------+--------------------------------------+
| id                | c6cc7f24-924b-4474-9276-a7a4d7b1448c |
| ip_protocol       | tcp                                  |
| ip_range          | 0.0.0.0/0                            |
| parent_group_id   | 96dd7907-f70e-4973-bf79-623fb6fe4e1e |
| port_range        | 22:22                                |
| remote_security_group |                                  |
+-------------------+--------------------------------------+
```

12.2.3 创建云主机

1. 生效 demo 用户环境变量

#. demo-openrc

2. 查看可用云主机类型

 # openstack flavor list

```
[root@controller ~]# openstack flavor list
+----+-----------+-------+------+-----------+-------+-----------+
| ID | Name      | RAM   | Disk | Ephemeral | VCPUs | Is Public |
+----+-----------+-------+------+-----------+-------+-----------+
| 1  | m1.tiny   | 512   | 1    | 0         | 1     | True      |
| 2  | m1.small  | 2048  | 20   | 0         | 1     | True      |
| 3  | m1.medium | 4096  | 40   | 0         | 2     | True      |
| 4  | m1.large  | 8192  | 80   | 0         | 4     | True      |
| 5  | m1.xlarge | 16384 | 160  | 0         | 8     | True      |
+----+-----------+-------+------+-----------+-------+-----------+
```

3. 查看可用镜像

 # openstack image list

```
[root@controller ~]#  openstack image list
+--------------------------------------+--------+--------+
| ID                                   | Name   | Status |
+--------------------------------------+--------+--------+
| ead0b366-5a9a-437e-892e-a0b4b40a7b3a | cirros | active |
+--------------------------------------+--------+--------+
```

4. 查看可用网络列表

 # openstack network list

```
[root@controller ~]#  openstack network list
+--------------------------------------+-------------+--------------------------------------+
| ID                                   | Name        | Subnets                              |
+--------------------------------------+-------------+--------------------------------------+
| 18a59b48-85d4-4b31-9e11-d466f5341c5b | provider    | e0e64a5d-e07a-4db5-87a5-7b59e2611d04 |
| 8f82ab7d-0cd5-4c79-adc4-e4365fa6ff11 | selfservice | 040a3afc-6b8b-49f5-af09-72fe990924bd |
+--------------------------------------+-------------+--------------------------------------+
```

5. 查看可用安全组

 # openstack security group list

```
[root@controller ~]# openstack security group list
+--------------------------------------+---------+------------------------+----------------------------------+
| ID                                   | Name    | Description            | Project                          |
+--------------------------------------+---------+------------------------+----------------------------------+
| 96dd7907-f70e-4973-bf79-623fb6fe4e1e | default | Default security group | 622d72c12b9b4fb49b76860b3178f490 |
+--------------------------------------+---------+------------------------+----------------------------------+
```

6. 创建云主机

 # openstack server create -flavor m1.tiny --image cirros --nic net-id=8f82ab7d-0cd5-4c79- adc4-e4365fa6ff11 --security-group default --key-name mykey selfservice-instance

```
[root@controller ~]# openstack server create --flavor m1.tiny --image cirros \
> --nic net-id=8f82ab7d-0cd5-4c79-adc4-e4365fa6ff11 --security-group default \
> --key-name mykey selfservice-instance
+--------------------------------------+------------------------------------------------+
| Field                                | Value                                          |
+--------------------------------------+------------------------------------------------+
| OS-DCF:diskConfig                    | MANUAL                                         |
| OS-EXT-AZ:availability_zone          |                                                |
| OS-EXT-STS:power_state               | 0                                              |
| OS-EXT-STS:task_state                | scheduling                                     |
| OS-EXT-STS:vm_state                  | building                                       |
| OS-SRV-USG:launched_at               | None                                           |
| OS-SRV-USG:terminated_at             | None                                           |
| accessIPv4                           |                                                |
| accessIPv6                           |                                                |
| addresses                            |                                                |
| adminPass                            | Z8BE9fKgEJDx                                   |
| config_drive                         |                                                |
| created                              | 2016-08-25T11:28:53Z                           |
| flavor                               | m1.tiny (1)                                    |
| hostId                               |                                                |
| id                                   | c814db96-ee55-464a-ac2e-d3ba1c0f7ef3           |
| image                                | cirros (ead0b366-5a9a-437e-892e-a0b4b40a7b3a)  |
| key_name                             | mykey                                          |
| name                                 | selfservice-instance                           |
| os-extended-volumes:volumes_attached | []                                             |
| progress                             | 0                                              |
| project_id                           | 622d72c12b9b4fb49b76860b3178f490               |
| properties                           |                                                |
| security_groups                      | [{u'name': u'default'}]                        |
| status                               | BUILD                                          |
| updated                              | 2016-08-25T11:28:55Z                           |
| user_id                              | 03a2b22588e74452ae3b2c1b803c1e6a               |
+--------------------------------------+------------------------------------------------+
```

注：

--nic net-id 选项：id 为 openstack network list 列表中的 selfservice 网络的 id。

创建的云主机的 id 为随机生成且唯一。

7. 查看云主机

 # openstack server list

```
[root@controller ~]# openstack server list
+--------------------------------------+----------------------+--------+---------------------+
| ID                                   | Name                 | Status | Networks            |
+--------------------------------------+----------------------+--------+---------------------+
| c814db96-ee55-464a-ac2e-d3ba1c0f7ef3 | selfservice-instance | ACTIVE | selfservice=10.0.0.3|
+--------------------------------------+----------------------+--------+---------------------+
```

8. 远程访问云主机

创建 provider 网络浮动 IP 地址：

 # openstack ip floating create provider

```
[root@controller ~]# openstack ip floating create provider
+-------------+--------------------------------------+
| Field       | Value                                |
+-------------+--------------------------------------+
| fixed_ip    | None                                 |
| id          | 85387ae1-af3f-4f13-9a6c-9303b83ac18a |
| instance_id | None                                 |
| ip          | 192.168.200.102                      |
| pool        | provider                             |
+-------------+--------------------------------------+
```

云主机与浮动 IP 地址关联：

 # openstack ip floating add 192.168.200.102 selfservice-instance

查看云主机：

 # openstack server list

```
[root@controller ~]# openstack server list
+--------------------------------------+-------------------+--------+---------------------------------+
| ID                                   | Name              | Status | Networks                        |
+--------------------------------------+-------------------+--------+---------------------------------+
| c814db96-ee55-464a-ac2e-d3ba1c0f7ef3 | selfservice-instance | ACTIVE | selfservice=10.0.0.3, 192.168.200.102 |
+--------------------------------------+-------------------+--------+---------------------------------+
```

验证并远程登录：

[root@compute ~]#**ping -c 4 192.168.200.102**
ssh cirros@192.168.200.102

```
[root@compute ~]# ssh cirros@192.168.200.102
The authenticity of host '192.168.200.102 (192.168.200.102)' can't be established.
RSA key fingerprint is 5c:d2:79:29:f8:11:9e:6c:85:15:4c:01:05:02:0a:a2.
Are you sure you want to continue connecting (yes/no)? yes
Warning: Permanently added '192.168.200.102' (RSA) to the list of known hosts.
cirros@192.168.200.102's password:

       [root@compute ~]# ping -c 4 192.168.200.102
       PING 192.168.200.102 (192.168.200.102) 56(84) bytes of data.
       64 bytes from 192.168.200.102: icmp_seq=1 ttl=63 time=1.51 ms
       64 bytes from 192.168.200.102: icmp_seq=2 ttl=63 time=0.739 ms
       64 bytes from 192.168.200.102: icmp_seq=3 ttl=63 time=0.584 ms
       64 bytes from 192.168.200.102: icmp_seq=4 ttl=63 time=0.625 ms

       --- 192.168.200.102 ping statistics ---
       4 packets transmitted, 4 received, 0% packet loss, time 3003ms
       rtt min/avg/max/mdev = 0.584/0.864/1.511/0.379 ms

       $
       $ ifconfig
       eth0      Link encap:Ethernet  HWaddr FA:16:3E:84:3C:B3
                 inet addr:10.0.0.3  Bcast:10.0.0.255  Mask:255.255.255.0
                 inet6 addr: fe80::f816:3eff:fe84:3cb3/64 Scope:Link
                 UP BROADCAST RUNNING MULTICAST  MTU:1450  Metric:1
                 RX packets:180 errors:0 dropped:0 overruns:0 frame:0
                 TX packets:196 errors:0 dropped:0 overruns:0 carrier:0
                 collisions:0 txqueuelen:1000
                 RX bytes:24571 (23.9 KiB)  TX bytes:21369 (20.8 KiB)

       lo        Link encap:Local Loopback
                 inet addr:127.0.0.1  Mask:255.0.0.0
                 inet6 addr: ::1/128 Scope:Host
                 UP LOOPBACK RUNNING  MTU:16436  Metric:1
                 RX packets:0 errors:0 dropped:0 overruns:0 frame:0
                 TX packets:0 errors:0 dropped:0 overruns:0 carrier:0
                 collisions:0 txqueuelen:0
                 RX bytes:0 (0.0 B)  TX bytes:0 (0.0 B)

       $ exit
Connection to 192.168.200.102 closed.
```

注：远程登录账户为 cirros，密码为 cubswin。

ifconfig 查看的 IP 地址 10.0.0.3 为云主机 IP 地址，192.168.200.102 为远程登录 IP 地址。

课后习题

1. 尝试使用 OpenStack 的 Web 界面运行一个云主机。
2. 使用命令运行一个云主机。

第13章

OpenStack 典型架构实现

13.1 OpenStack 架构及规划

本章将实现一种典型云平台架构（如图 13-1 所示）。

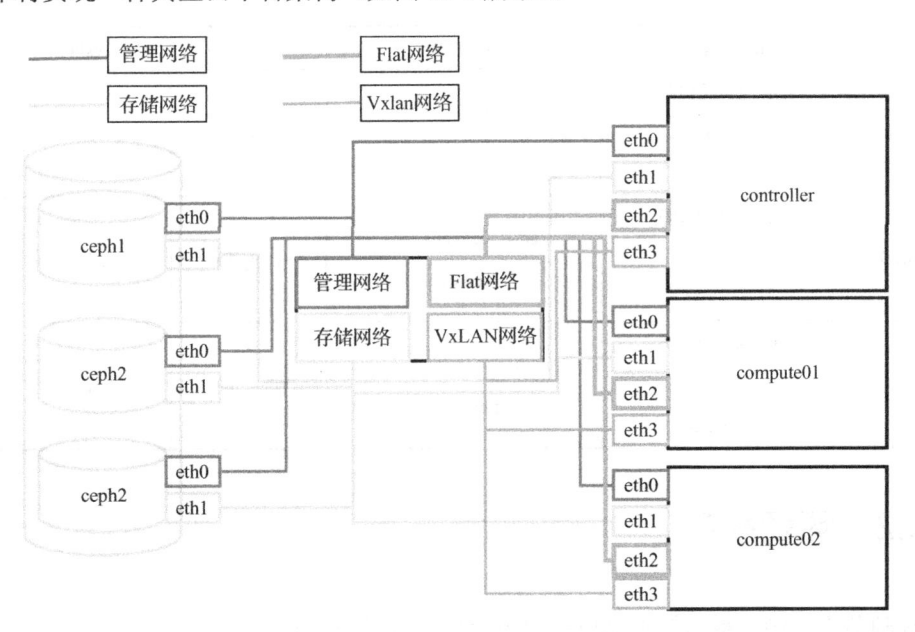

图 13-1 典型云平台架构

该架构采用 1+2+3 模式，即 1 个控制节点+2 个计算节点+3 个存储节点。存储为 Ceph 存储，设置副本数为 3，实现存储的高可用性；2 个计算节点实现虚拟机的迁移、备份；1 个控制节点主要部署 OpenStack 相关管理服务，实现整个平台的管理。存储服务器共有 2 个网络，计算节点和控制节点各有 4 个网卡。在该架构中将管理网络、存储网络、Flat 网络、VxLAN 网络隔离，不同的流量走不同的网络，防止网络流量过大。这里要注意的是，OpenStack 如果选用 VLAN 网络，VLAN 网络的物理网卡所接的交换机接口需要配成 trunk 模式。

整个架构由存储平台和 OpenStack 平台构成的信息规划如下。

Ceph 环境信息如表 13-1 所示。

表 13-1 Ceph 环境信息

Hostname	eth0（管理）	eth1（存储）	OS
ceph1	ip:10.78.70.107 gt:10.78.70.1 prefix: 24	ip: 10.78.201.147 prefix: 24	CentOS 7.2
ceph2	ip:10.78.70.108 gt:10.78.70.1 prefix: 24	ip: 10.78.201.148 prefix: 24	CentOS 7.2
ceph3	ip:10.78.70.109 gt:10.78.70.1 prefix: 24	ip: 10.78.201.149 prefix: 24s	CentOS 7.2

OpenStack 平台信息如表 13-2 所示。

表 13-2 OpenStack 平台信息

Hostname	eth0（管理）	eth1（存储）	eth2（flat）	eth3（vxlan）	OS
controller	ip:10.78.70.104 gt:10.78.70.1 prefix:24	eth1:10.78.201.143 prefix:24	vlan_flat	trunk	CentOS 7.2
compute 01	ip:10.78.70.105 gt:10.78.70.1 prefix:24	eth1:10.78.201.145 prefix:24	vlan_flat	trunk	CentOS 7.2
compute 02	ip:10.78.70.106 gt:10.78.70.1 prefix:24	eth1:10.78.201.146 prefix:24	vlan_flat	trunk	CentOS 7.2

13.2 环境准备

以控制节点为例，根据规划完成各个节点的环境准备工作。

1. 设置主机名

```
# echo controller> /etc/hostname
# hostname controller
```

2. 配置主机名映射编辑/etc/hosts 文件并添加如下内容

```
# vi /etc/hosts
10.78.201.147    ceph1
10.78.201.148    ceph2
10.78.201.149    ceph3
10.78.70.104     controller
10.78.70.105     compute01
10.78.70.106     compute02
```

3. 关闭防火墙

```
# service firewalld stop
# chkconfig firewalld off
# sed -i 's/SELINUX=enforcing/SELINUX=disabled/g' /etc/selinux/config
# setenforce 0
```

4. 配置 yum 源

这里主要添加 Ceph 的源，对其他源在前面章节已做过配置。

```
vim    /etc/yum.repo.d/ceph.repo
[ceph]
name=Ceph noarch packages
baseurl=http://mirrors.163.com/ceph/rpm-hammer/el7/x86_64/
enabled=1
gpgcheck=1
type=rpm-md
gpgkey=http://mirrors.163.com/ceph/keys/release.asc
yum clean all && yum makecache
```

5. 安装基础软件包

yum install -y python-pip vim-enhanced bash-completion patch net-tools yum-plugin-priorities lsof openssh-clientstree lrzsz sysstat which nload screen wget lsof ntp telnet unzip axel supervisor libsysfs-devel libsysfs usbutils automake autoconf libtool glib-devel glibc-commonglib glib2 glib2-devel glib2-doc.noarch glibc-devel glibc-static kernel-devel kernel kernel-headers dbus-python-devel dbus-devel dbus-glib-devel dbus-libs pygobject3-devel pygobject2 fping python-pip python-tornado python-rados ntpdate firewalld python-devel gcc gcc-c++ sg3_utilspython-openstackclient

注：这些软件包中的一些不是 OpenStack 相关的软件包，更多的是运维用到的相关工具。比如，我们在检查网络流量时常用的工具 nload。

6. 时间同步

```
# yum install ntp -y
```

编辑**/etc/ntp.conf** 文件。
添加以下内容：

```
server 127.127.1.0
fudge 127.127.1.0 stratum 10
```

启动 ntp 服务并设置开机自启动：

```
# systemctl start ntpd.service
# systemctl enable ntpd.service
# ntpstat
```

这一步是为了查看状态,直到显示第二次的结果即成功。

其他节点同步时间参考前文 ntp 服务准备工作。只需要执行 ntpdate controller。

7. 设置 limits

```
# vi /etc/security/limits.conf
# *    soft       nofile    65535
# *    hard       nofile    65535
```

8. 设置/etc/sysctl.conf

```
# echo "net.ipv4.conf.all.rp_filter = 0">> /etc/sysctl.conf
# echo "net.ipv4.conf.default.rp_filter = 0">> /etc/sysctl.conf
# echo "net.ipv4.ip_forward = 1">> /etc/sysctl.conf
# echo "net.bridge.bridge-nf-call-iptables = 1">> /etc/sysctl.conf
# echo "net.bridge.bridge-nf-call-ip6tables = 1">> /etc/sysctl.conf
# echo "net.ipv4.ip_nonlocal_bind = 1">> /etc/sysctl.conf
# echo "net.ipv4.conf.default.accept_source_route = 0">> /etc/sysctl.conf
# sysctl -p
```

13.3 Ceph 集群部署及配置

这里选择 Ceph 作为 OpenStack 的云存储,将云平台的镜像、虚拟机等数据存入 Ceph 中。

13.3.1 Ceph 的相关知识

Ceph:一个 Linux PB 级分布式文件系统。

Linux 持续不断进军可扩展计算空间,特别是可扩展存储空间。Ceph 最近才加入 Linux 中令人印象深刻的文件系统备选行列。它是一个分布式文件系统,能够在维护 POSIX 兼容性的同时加入复制和容错功能。探索 Ceph 的架构,学习它如何提供容错功能,简化海量数据管理。

Ceph 最初是一项关于存储系统的 PhD 研究项目,由 Sage Weil 在 University of California, Santa Cruz(UCSC)实施。但是到了 2010 年 3 月底,可以在主线 Linux 内核(从 2.6.34 版开始)中找到 Ceph 的身影。虽然 Ceph 可能还不适用于生产环境,但它对测试目的还是非常有用的。下面探讨 Ceph 文件系统及其独有的功能,这些功能让它成为可扩展分布式存储的最有吸引力的备选。

13.3.2 Ceph 目标

开发一个分布式文件系统需要多方努力,但是如果能准确地解决问题,它就是无价的。Ceph 的目标简单地定义为:

- 可轻松扩展到数 PB 容量;

- 对多种工作负载的高性能（每秒输入/输出操作[IOPS]和带宽）；
- 高可靠性。

不幸的是，这些目标之间会互相竞争（例如，可扩展性会降低或者抑制性能，或者影响可靠性）。Ceph 开发了一些非常有趣的概念（例如，动态元数据分区，数据分布和复制），对这些概念在这里只进行简短的探讨。Ceph 的设计还包括保护单一点故障的容错功能，它假设大规模（PB 级存储）存储故障是常见现象而不是例外情况。最后，它的设计并没有假设某种特殊工作负载，但是包括适应变化的工作负载，提供最佳性能的能力。它利用 POSIX 的兼容性完成所有这些任务，允许它对当前依赖 POSIX 语义（通过以 Ceph 为目标的改进）的应用进行透明的部署。最后，Ceph 是开源分布式存储，也是主线 Linux 内核（2.6.34）的一部分。

13.3.3　Ceph 架构

下面探讨 Ceph 的架构以及高端的核心要素。然后拓展到另一层次，说明 Ceph 中一些关键的方面，进行更详细的探讨。

Ceph 生态系统可以大致划分为以下 4 个部分（见图 13-2）。

（1）Clients：客户端（数据用户）。

（2）Metadata server cluster：元数据服务器（缓存和同步分布式元数据）。

（3）Object storage cluster：对象存储集群（将数据和元数据作为对象存储，执行其他关键职能）。

（4）Cluster monitors：集群监视器（执行监视功能）。

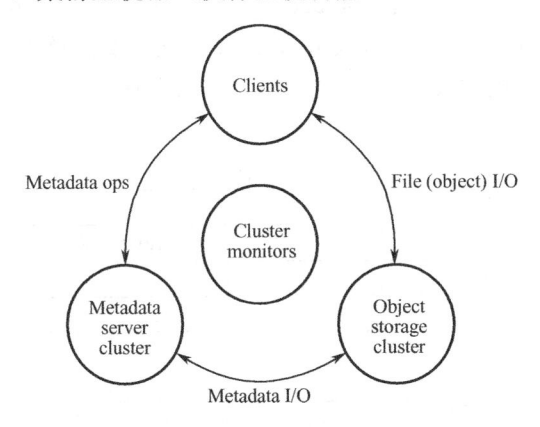

图 13-2　Ceph 生态系统的概念架构

如图 13-2 所示，客户使用元数据服务器，执行元数据操作（来确定数据位置）。元数据服务器管理数据位置，以及在何处存储新数据。值得注意的是，元数据存储在一个存储集群（标为"元数据 I/O"）中。实际的文件 I/O 发生在客户和对象存储集群之间。这样，更高层次的 POSIX 功能（例如打开、关闭、重命名）就由元数据服务器管理，但 POSIX 功能（例如读和写）则直接由对象存储集群管理。

Ceph 生态系统简化后的分层视图如图 13-3 所示。一系列服务器通过一个客户界面访问 Ceph 生态系统，这就清楚地表示了元数据服务器和对象级存储器之间的关系。分布式存储系统可以在一些层中查看，包括一个存储设备的格式[Extent and B-tree-based Object File System

云操作系统应用（OpenStack）

(EBOFS)或者一个备选]，还有一个设计用于管理数据复制、故障检测、恢复，以及随后的数据迁移的覆盖管理层，称为 Reliable Autonomic Distributed Object Storage（RADOS）。最后，监视器用于识别组件故障，包括随后的通知。

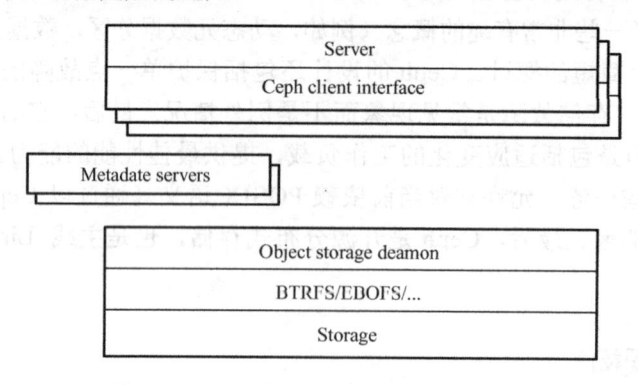

图 13-3　Ceph 生态系统简化后的分层视图

13.3.4　Ceph 组件

在了解了 Ceph 的概念架构之后，便可以挖掘到另一个层次，了解在 Ceph 中实现的主要组件。Ceph 和传统的文件系统之间的重要差异之一是，它将智能都用在了生态环境而不是文件系统本身。

图 13-4 显示了一个简单的 Ceph 生态系统。Ceph client 是 Ceph 文件系统的用户。Ceph metadata daemon 提供元数据服务器，而 ceph object storage daemon 提供实际存储（对数据和元数据两者）。最后，Ceph monitor 提供了集群管理。要注意的是，Ceph 客户、对象存储端点、元数据服务器（根据文件系统的容量）可以有许多，而且至少有一对冗余的监视器。

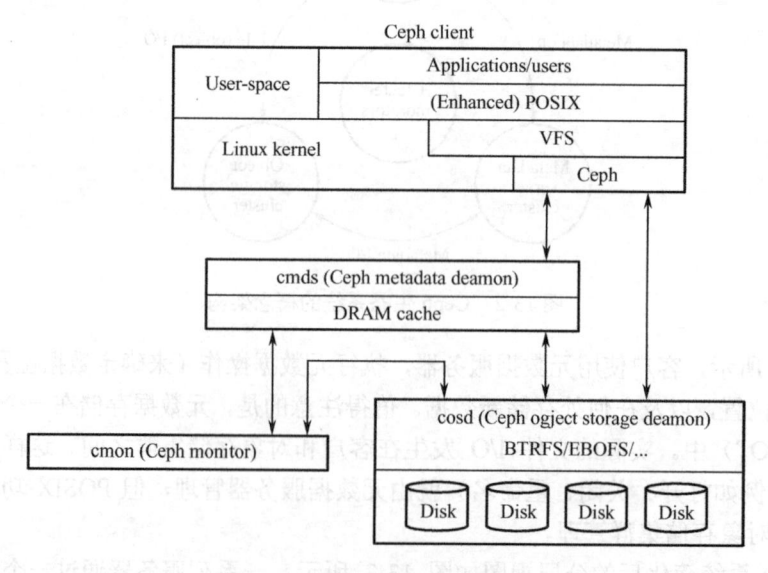

图 13-4　简单的 Ceph 生态系统

1. Ceph 客户端

因为 Linux 显示文件系统的一个公共界面[通过虚拟文件系统交换机（VFS）]，Ceph 的用户透视图就是透明的。管理员的透视图肯定是不同的，考虑到很多服务器会包含存储系统这一潜在因素。从用户的角度看，他们访问大容量的存储系统，却不知道下面聚合成一个大容量的存储池的元数据服务器、监视器，还有独立的对象存储设备。用户只是简单地看到一个安装点，在这点上可以执行标准文件 I/O。

Ceph 文件系统——或者至少是客户端接口——在 Linux 内核中实现。值得注意的是，在大多数文件系统中，所有的控制和智能在内核的文件系统源本身中执行。但是，在 Ceph 中，文件系统的智能分布在节点上，这简化了客户端接口，并为 Ceph 提供了大规模（甚至动态）扩展能力。

Ceph 使用一个有趣的备选，而不是依赖分配列表（将磁盘上的块映射到指定文件的元数据）。Linux 透视图中的一个文件会分配到一个来自元数据服务器的 inode number（INO），对于文件这是一个唯一的标识符。然后，文件被推入一些对象中（根据文件的大小）。使用 INO 和 object number（ONO），每个对象都分配到一个对象 ID（OID）。在 OID 上使用一个简单的哈希，每个对象都被分配到一个放置组。放置组（标识为 PGID）是一个对象的概念容器。最后，放置组到对象存储设备的映射是一个伪随机映射，使用一个称为 Controlled Replication Under Scalable Hashing（CRUSH）的算法。这样，放置组（以及副本）到存储设备的映射就不用依赖任何元数据，而是依赖一个伪随机的映射函数。这种操作是理想的，因为它把存储的开销最小化，简化了分配和数据查询。

分配的最后组件是集群映射。集群映射是设备的有效表示，显示了存储集群。有了 PGID 和集群映射，就可以定位任何对象。

2. Ceph 元数据服务器

元数据服务器（cmds）的工作就是管理文件系统的名称空间。虽然元数据和数据两者都存储在对象存储集群中，但两者分别管理，支持可扩展性。事实上，元数据在一个元数据服务器集群中被进一步拆分，元数据服务器能够自适应地复制和分配名称空间，避免出现热点。如图 13-5 所示，元数据服务器管理名称空间部分，可以（为冗余和性能）进行重叠。元数据服务器到名称空间的映射在 Ceph 中使用动态子树逻辑分区执行，它允许 Ceph 对变化的工作负载进行调整（在元数据服务器之间迁移名称空间），同时保留性能的位置。

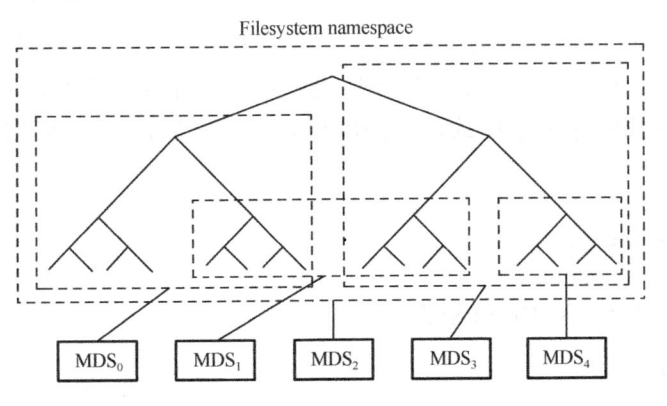

图 13-5 元数据服务器的 Ceph 名称空间的分区

但是，因为每个元数据服务器只是简单地管理客户端入口的名称空间，它的主要应用就是一个智能元数据缓存（因为实际的元数据最终存储在对象存储集群中）。进行写操作的元数据被缓存在一个短期的日志中，它最终还是被推入物理存储器中。这个动作允许元数据服务器将最近的元数据回馈给客户（这在元数据操作中很常见）。这个日志对故障恢复也很有用：如果元数据服务器发生故障，它的日志就会被重放，保证元数据安全存储在磁盘上。

元数据服务器管理 inode 空间，将文件名转变为元数据。元数据服务器将文件名转变为索引节点、文件大小和 Ceph 客户端用于文件 I/O 的分段数据（布局）。

3. Ceph 监视器

Ceph 包含实施集群映射管理的监视器，但是故障管理的一些要素是在对象存储本身中执行的。当对象存储设备发生故障或者添加新设备时，监视器就检测和维护一个有效的集群映射。这个功能按一种分布的方式执行，这种方式中映射升级可以和当前的流量通信。Ceph 使用 Paxos，它是一系列分布式共识算法。

4. Ceph 对象存储

和传统的对象存储类似，Ceph 存储节点不仅包括存储，还包括智能。传统的驱动是只响应来自启动者的命令的简单目标。但是，对象存储设备是智能设备，它能作为目标和启动者，支持与其他对象存储设备的通信和合作。

从存储角度来看，Ceph 对象存储设备执行从对象到块的映射（在客户端的文件系统层中常常执行的任务）。这个动作允许本地实体以最佳方式决定怎样存储一个对象。Ceph 的早期版本在一个名为 EBOFS 的本地存储器上实现一个自定义低级文件系统。这个系统实现一个到底层存储的非标准接口，这个底层存储已针对对象语义和其他特性（例如对磁盘提交的异步通知）调优。今天，B-tree 文件系统（BTRFS）可以被用于存储节点，它已经实现了部分必要功能（例如嵌入式完整性）。

因为 Ceph 客户实现 CRUSH，而且对磁盘上的文件映射块一无所知，所以下面的存储设备就能安全地管理对象到块的映射。这允许存储节点复制数据（当发现一个设备出现故障时）。分配故障恢复也允许存储系统扩展，因为故障检测和恢复跨生态系统分配。Ceph 称其为 RADOS。

13.3.5 Ceph 的地位和未来

虽然 Ceph 现在被集成在主线 Linux 内核中，但只是标识为实验性的。在这种状态下的文件系统对测试是有用的，但是对生产环境没有做好准备。考虑到 Ceph 加入到 Linux 内核的行列，还有其创建人想继续研发的动机，不久之后，它应该能用于满足海量存储需要。

13.3.6 Ceph 的搭建

1. 在 Ceph 节点上安装 Ceph 相关软件包

```
# yum install ceph
```

注：如果 ceph-deploy 安装不上，可以到 http://download.ceph.com/rpm-giant/rhel7/ 下载 ceph-deploy 的安装包，通过 yum 直接安装。

2. Ceph 各个节点直接配置 SSH 互信

以 ceph1 为例：

```
# ssh-keygen -t dsa -P '' -f ~/.ssh/id_dsa
# cat ~/.ssh/id_dsa.pub |ssh root@ceph1"cat - >> ~/.ssh/authorized_keys"
# cat ~/.ssh/id_dsa.pub |ssh root@ceph2"cat - >> ~/.ssh/authorized_keys"
# cat ~/.ssh/id_dsa.pub |ssh root@ceph3"cat - >> ~/.ssh/authorized_keys"
```

3. 在 ceph1 上安装 ceph-deploy

通过 ceph-deploy 部署 ceph 集群，每个存储节点有三块硬盘作为 ceph 的 osd。

```
# yum install ceph-deploy -y
# mkdir -p /root/ceph-management
# cd /root/ceph-management
# ceph-deploy new    ceph1
# ceph-deploy mon create ceph1 ceph2 ceph3
# ceph-deploy gatherkeys ceph1 ceph2 ceph3
# ceph-deploy disk zap controller:sd{b,c,d}
# ceph-deploy --overwrite-conf osd prepare ceph1:/dev/sdb
# ceph-deploy --overwrite-conf osd prepare ceph1:/dev/sdc
# ceph-deploy --overwrite-conf osd prepare ceph1:/dev/sdd
# ceph-deploy --overwrite-conf osd prepare ceph2:/dev/sdb
# ceph-deploy --overwrite-conf osd prepare ceph2:/dev/sdc
# ceph-deploy --overwrite-conf osd prepare ceph2:/dev/sdd
# ceph-deploy --overwrite-conf osd prepare ceph3:/dev/sdb
# ceph-deploy --overwrite-conf osd prepare ceph3:/dev/sdc
# ceph-deploy --overwrite-conf osd prepare ceph3:/dev/sdd
# ceph-deploy osd activate ceph1:/dev/sdb
# ceph-deploy osd activate ceph1:/dev/sdc
# ceph-deploy osd activate ceph1:/dev/sdd
# ceph-deploy osd activate ceph2:/dev/sdb
# ceph-deploy osd activate ceph2:/dev/sdc
# ceph-deploy osd activate ceph2:/dev/sdd
# ceph-deploy osd activate ceph3:/dev/sdb
# ceph-deploy osd activate ceph3:/dev/sdc
# ceph-deploy osd activate ceph3:/dev/sdd
```

每个存储节点：设置做 osd 的磁盘开机后自动挂载。

```
# df -h |grep ceph |awk '{print "mount",$1,$NF}' >> /etc/rc.local
```

4. 配置云平台资源池

```
# ceph osd pool create volumes 512
# ceph osd pool create images 512
# ceph osd pool set volumes size 3
# ceph osd pool set images size 3
```

为后面 OpenStack Cinder 服务和 Glance 服务做准备，创建 Images 池和 Volumes 池分别作为 Glance 和 Cinder 服务的存储池。

13.4 OpenStack 搭建

13.4.1 安装数据库

1. 在控制节点安装 Mariadb 服务，作为 OpenStack 服务的数据库存储

```
# yum install mariadb mariadb-server python2-PyMySQL
```

2. 修改数据库配置文件

```
# vim /etc/my.cnf.d/openstack.cnf
```

在[**mysqld**]部分，做如下添加。

```
bind-address = 10.78.70.104
default-storage-engine = innodb
innodb_file_per_table
collation-server = utf8_general_ci
init-connect = 'SET NAMES utf8'
character-set-server = utf8
open_files_limit = 165535
max_connections = 1024
```

3. 启动数据库服务和配置开机自启动

```
# systemctl enable mariadb.service
# systemctl start mariadb.service
```

4. 配置 MySQL 的密码

```
# mysql_secure_installation
```

13.4.2 安装消息队列服务

1. 在控制节点安装 Rabbitmq，作为消息队列服务

```
# yum install rabbitmq-server
```

2. 启动队列服务并设置开机自启动

```
# systemctl enable rabbitmq-server.service
# systemctl start rabbitmq-server.service
```

3. 创建用户和密码、权限

```
# rabbitmqctl add_user openstack $RABBIT_PASS
# rabbitmqctl set_permissions openstack ".*" ".*" ".*"
```

13.4.3 安装 Memcached 服务

1. 在控制节点安装 Memcached 软件包

\# yum install memcached python-memcached

2. 启动 Memcached 服务并设置开机自启动

\# systemctl enable memcached.service
\# systemctl start memcached.service

13.4.4 安装认证服务

1. 创建 Keystone 数据库并配置 keystone 的访问权限

\# mysql -uroot -phuayun
\# CREATE DATABASE keystone;
\# GRANT ALL PRIVILEGES ON keystone.* TO 'keystone'@'localhost' IDENTIFIED BY 'KEYSTONE_DBPASS';
\# GRANT ALL PRIVILEGES ON keystone.* TO 'keystone'@'%' IDENTIFIED BY 'KEYSTONE_DBPASS';
\# GRANT ALL PRIVILEGES ON keystone.* TO 'keystone'@'controller' IDENTIFIED BY 'KEYSTONE_DBPASS';

2. 安装 Keystone 软件包

\# yum install openstack-keystone httpd mod_wsgi

3. 修改 Keystone 配置文件

[DEFAULT]
admin_token = ADMIN_TOKEN
　[database]
connection =mysql+pymysql://keystone:KEYSTONE_DBPASS@controller/keystone
max_pool_size=70
max_overflow=30
　[token]
expiration=36000
provider = fernet

4. 初始化数据库

\# su -s /bin/sh -c "keystone-manage db_sync" keystone

5. 初始化 Fernet keys

\# keystone-manage fernet_setup --keystone-user keystone --keystone-group keystone

6. 配置 Apache Http Server

编辑配置文件：

\# vim /etc/httpd/conf/httpd.conf

ServerName controller:80

创建配置文件

```
# vim /etc/httpd/conf.d/wsgi-keystone.conf
Listen 5000
Listen 35357

<VirtualHost *:5000>
    WSGIDaemonProcess keystone-public processes=5 threads=1 user=keystone group=keystone display-name=%{GROUP}
    WSGIProcessGroup keystone-public
    WSGIScriptAlias / /usr/bin/keystone-wsgi-public
    WSGIApplicationGroup %{GLOBAL}
    WSGIPassAuthorization On
    ErrorLogFormat "%{cu}t %M"
    ErrorLog /var/log/httpd/keystone-error.log
    CustomLog /var/log/httpd/keystone-access.log combined

    <Directory /usr/bin>
        Require all granted
    </Directory>
</VirtualHost>

<VirtualHost *:35357>
    WSGIDaemonProcess keystone-admin processes=5 threads=1 user=keystone group=keystone display-name=%{GROUP}
    WSGIProcessGroup keystone-admin
    WSGIScriptAlias / /usr/bin/keystone-wsgi-admin
    WSGIApplicationGroup %{GLOBAL}
    WSGIPassAuthorization On
    ErrorLogFormat "%{cu}t %M"
    ErrorLog /var/log/httpd/keystone-error.log
    CustomLog /var/log/httpd/keystone-access.log combined

    <Directory /usr/bin>
        Require all granted
    </Directory>
</VirtualHost>
```

7. 启动服务并设置开机自启动

```
# systemctl enable httpd.service
# systemctl start httpd.service
```

8. 创建租户和角色

```
# export OS_TOKEN=ADMIN_TOKEN
# export OS_URL=http://10.78.70.104:35357/v3
# export OS_IDENTITY_API_VERSION=3
```

9. 创建认证服务 API

\# openstack service create --name keystone --description "OpenStack Identity" identity
\# openstack endpoint create --region RegionOne identity publichttp://10.78.70.104:5000/v3
\# openstack endpoint create --region RegionOne identity internal http://10.78.70.104:5000/v3
\# openstack endpoint create --region RegionOne identity admin http://10.78.70.104:35357/v3

10. 创建 default 域

\# openstack domain create --description "Default Domain" default

11. 创建 admin 项目

\# openstack project create --domain default --description "Admin Project" admin

12. 创建 admin 用户

\# openstack user create --domain default --password ADMIN_PASS admin

13. 创建 admin 角色

\# openstack role create admin

14. 创建 admin 项目、角色、用户关联

\# openstack role add --project admin --user admin admin

15. 创建 Service Project

\# openstack project create --domain default --description "Service Project" service

16. 创建 admin-openrc

\# vim admin-openrc
export OS_PROJECT_DOMAIN_NAME=default
export OS_USER_DOMAIN_NAME=default
export OS_PROJECT_NAME=admin
export OS_USERNAME=admin
export OS_PASSWORD=ADMIN_PASS
export OS_AUTH_URL=http://10.78.70.104:35357/v3
export OS_IDENTITY_API_VERSION=3
export OS_IMAGE_API_VERSION=2

验证：

\# source admin-openrc
\# openstack token issue

此处应该有 expires、id project_id、user_id 的返回。

13.4.5 安装镜像服务

1. 创建 Glance 数据库并配置权限

\# mysql -u root -p
CREATE DATABASE glance;

GRANT ALL PRIVILEGES ON glance.* TO 'glance'@'localhost' IDENTIFIED BY 'GLANCE_DBPASS';

GRANT ALL PRIVILEGES ON glance.* TO 'glance'@'%' IDENTIFIED BY 'GLANCE_DBPASS';

GRANT ALL PRIVILEGES ON glance.* TO 'glance'@'controller' IDENTIFIED BY 'GLANCE_DBPASS';

2. 创建 Glance 服务的认证

source admin-openrc

创建用户：

openstack user create --domain default --password GLANCE_PASS glance

创建角色：

openstack role add --project service --user glance admin

创建 service entity 和 api endpoints：

openstack service create --name glance \
    --description "OpenStack Image" image
openstack endpoint create --region RegionOne \
    image public http://10.78.70.104:9292
openstack endpoint create --region RegionOne \
    image internal http://10.78.70.104:9292
openstack endpoint create --region RegionOne \
    image admin http://10.78.70.104:9292

3. 安装配置 Glance 服务

Glance 服务选择 ceph 作为后端存储：

yum install -y openstack-glance ceph-common

添加权限文件：

cd /etc/ceph/

将 ceph 节点 /etc/ceph 目录下的 ceph.conf nova-ceph.conf ceph.client.admin.keyring 文件复制到此目录。

4. 配置 Glance 服务

vim /etc/glance/glance-api.conf

1）示例 1

[DEFAULT]
show_image_direct_url = True
workers = 20
default_log_levels
=amqp=WARN,amqplib=WARN,boto=WARN,qpid=WARN,sqlalchemy=WARN,suds=INFO,oslo.messaging=INFO,iso8601=WARN,requests.packages.urllib3.connectionpool=WARN,urllib3.connectionpool=WARN,websocket=WARN,requests.packages.urllib3.util.retry=WARN,urllib3.util.retry=WARN,keystonemiddleware=WARN,routes.middleware=WARN,stevedore=WARN,taskflow=WARN,keystoneauth=WARN,oslo.cache=INFO,dogpile.core.dogpile=INFO

```
[cors]
[cors.subdomain]
[database]
connection = mysql+pymysql://glance:GLANCE_DBPASS@controller/glance
max_pool_size = 100
max_overflow = 30
[glance_store]
stores=rbd
default_store = rbd
rbd_store_chunk_size = 8
rbd_store_pool = images
rbd_store_user = admin
rbd_store_ceph_conf = /etc/ceph/ceph.conf
[image_format]
[keystone_authtoken]
auth_uri = http://10.78.70.104:5000
auth_url = http://10.78.70.104:35357
memcached_servers = controller:11211
auth_type = password
project_domain_name = default
user_domain_name = default
project_name = service
username = glance
password = ADMIN_PASS
[oslo_concurrency]
[oslo_messaging_amqp]
[oslo_messaging_notifications]
[oslo_messaging_rabbit]
[oslo_policy]
[paste_deploy]
flavor = keystone
[profiler]
[store_type_location_strategy]
[task]
[taskflow_executor]
```

 # vim /etc/glance/glance-registry.conf

2)示例2

```
[DEFAULT]
workers = 20
[database]
connection = mysql+pymysql://glance:GLANCE_DBPASS@controller/glance
[glance_store]
[keystone_authtoken]
auth_uri = http://10.78.70.104:5000
auth_url = http://10.78.70.104:35357
memcached_servers = controller:11211
```

```
auth_type = password
project_domain_name = default
user_domain_name = default
project_name = service
username = glance
password = ADMIN_PASS
[matchmaker_redis]
[oslo_messaging_amqp]
[oslo_messaging_notifications]
[oslo_messaging_rabbit]
[oslo_policy]
[paste_deploy]
flavor = keystone
[profiler]
```

5. 初始化数据库

```
# su -s /bin/sh -c "glance-manage db_sync" glance
```

6. 启动服务并设置开机自启动

```
# systemctl enable openstack-glance-api.service \
openstack-glance-registry.service
systemctl start systemctl start openstack-glance-api.service \
openstack-glance-registry.service
```

13.4.6 在控制节点安装 Cinder 服务

1. 创建 Cinder 数据库并配置 Cinder 数据库访问权限

```
# mysql -u root -p
CREATE DATABASE cinder;
GRANT ALL PRIVILEGES ON cinder.* TO 'cinder'@'localhost' IDENTIFIED BY 'CINDER_DBPASS';
GRANT ALL PRIVILEGES ON cinder.* TO 'cinder'@'%' IDENTIFIED BY 'CINDER_DBPASS';
GRANT ALL PRIVILEGES ON cinder.* TO 'cinder'@'controller' IDENTIFIED BY 'CINDER_DBPASS';
```

2. 在 Keystone 中创建 cinder 用户

在 Keystone 中创建 cinder 用户，并将 cinder 用户分配到 service 租户下给予 admin 的角色，创建 Cinder 服务的端点。

```
# source admin-openrc
# openstack user create --domain default --password CINDER_PASS cinder
# openstack role add --project service --user cinder admin
```

1）创建 service（v1 和 v2）

```
# openstack service create --name cinder \
    --description "OpenStack Block Storage" volume
# openstack service create --name cinderv2 \
```

--description "OpenStack Block Storage" volumev2

2）创建 endpoint（v1 和 v2）

openstack endpoint create --region RegionOne \
   volume public http://10.78.70.104:8776/v1/%\(tenant_id\)s
openstack endpoint create --region RegionOne \
   volume internal http://10.78.70.104:8776/v1/%\(tenant_id\)s

openstack endpoint create --region RegionOne \
   volume admin http://10.78.70.104:8776/v1/%\(tenant_id\)s

openstack endpoint create --region RegionOne \
   volumev2 public http://10.78.70.104:8776/v2/%\(tenant_id\)s

openstack endpoint create --region RegionOne \
   volumev2 internal http://10.78.70.104:8776/v2/%\(tenant_id\)s

openstack endpoint create --region RegionOne \
   volumev2 admin http://10.78.70.104:8776/v2/%\(tenant_id\)s

3. 安装软件包

yum install openstack-cinder

4. 修改 Cinder 配置文件

```
# vim /etc/cinder/cinder.conf
[DEFAULT]
my_ip = 10.78.70.104
glance_api_version = 2
auth_strategy = keystone
rbd_pool=volumes
rbd_user=admin
rbd_ceph_conf=/etc/ceph/ceph.conf
rbd_flatten_volume_from_snapshot = false
rbd_max_clone_depth = 5
rbd_store_chunk_size = 4
rados_connect_timeout = -1
quota_volumes = 1000
quota_snapshots = 100
volume_driver = cinder.volume.drivers.rbd.RBDDriver
rpc_backend = rabbit
[BACKEND]
[BRCD_FABRIC_EXAMPLE]
[CISCO_FABRIC_EXAMPLE]
[COORDINATION]
[FC-ZONE-MANAGER]
[KEYMGR]
[cors]
```

```
[cors.subdomain]
[database]
connection = mysql+pymysql://cinder:CINDER_DBPASS@controller/cinder
[keystone_authtoken]
auth_uri = http://10.78.70.104:5000
auth_url = http://10.78.70.104:35357
memcached_servers = controller:11211
auth_type = password
project_domain_name = default
user_domain_name = default
project_name = service
username = cinder
password = ADMIN_PASS
[matchmaker_redis]
[oslo_concurrency]
lock_path = /var/lib/cinder/tmp
[oslo_messaging_amqp]
[oslo_messaging_notifications]
driver = messagingv2
[oslo_messaging_rabbit]
rabbit_host = controller
rabbit_userid = openstack
rabbit_password = RABBIT_PASS
[oslo_middleware]
[oslo_policy]
[oslo_reports]
[oslo_versionedobjects]
[ssl]
```

5. 初始化数据库

su -s /bin/sh -c "cinder-manage db sync" cinder

6. 启动服务并设置开机自启动

systemctl enable openstack-cinder-api.service openstack-cinder-scheduler.service openstack-cinder-volume.service

systemctl restart openstack-cinder-api.service openstack-cinder-scheduler.service openstack-cinder-volume.service

7. 验证

cinder service-list
cinder create --display-name demo-volume1 1
cinder list

8. 调整 Volume 使 ceph 为 rdb 类型

1）修改配置

vim /etc/cinder/cinder.conf

```
[rbd]
volume_driver = cinder.volume.drivers.rbd.RBDDriver
# glance_api_version = 2
rbd_pool=volumes
rbd_user=admin
rbd_ceph_conf=/etc/ceph/ceph.conf
rbd_flatten_volume_from_snapshot = false
rbd_max_clone_depth = 5
rbd_store_chunk_size = 4
rados_connect_timeout = -1
volume_backend_name = RBD
```

2）执行命令

```
# cinder type-create rbd --is-public true
# cinder type-key rbd set volume_backend_name=RBD
```

3）检查配置

```
# cinder extra-specs-list
```

13.4.7 安装计算服务

1. 在控制节点安装 Nova 相关服务

1）创建 Nova 和 Nova_api 数据库并配置数据库访问权限

```
#mysql -u root –p
CREATE DATABASE nova_api;
CREATE DATABASE nova;
```

2）赋予权限

```
GRANT ALL PRIVILEGES ON nova_api.* TO 'nova'@'localhost' \
  IDENTIFIED BY 'NOVA_DBPASS';
GRANT ALL PRIVILEGES ON nova_api.* TO 'nova'@'%' \
  IDENTIFIED BY 'NOVA_DBPASS';
GRANT ALL PRIVILEGES ON nova.* TO 'nova'@'localhost' IDENTIFIED BY 'NOVA_DBPASS';
GRANT ALL PRIVILEGES ON nova.* TO 'nova'@'%' IDENTIFIED BY 'NOVA_DBPASS';
```

3）在 Keystone 中配置 Nova 服务用户、租户、角色、端点等信息

```
# source admin-openrc
# openstack user create --domain default \
  --password NOVA_PASS nova
openstack role add --project service --user nova admin
```

4）创建计算服务，并创建计算服务的 endpoint

```
# openstack service create --name nova \
  --description "OpenStack Compute" compute
# openstack endpoint create --region RegionOne \
  compute public http://10.78.70.104:8774/v2.1/%\(tenant_id\)s
```

```
# openstack endpoint create --region RegionOne \
  compute internal http://10.78.70.104:8774/v2.1/%\(tenant_id\)s
# openstack endpoint create --region RegionOne \
  compute admin http://10.78.70.104:8774/v2.1/%\(tenant_id\)s
```

5）在控制节点安装 Nova 相关软件包

```
# yum install openstack-nova-api openstack-nova-cert openstack-nova-conductor openstack-nova-console openstack-nova-novncproxy openstack-nova-scheduler
```

6）在控制节点配置 nova.conf 文件

```
# vim /etc/nova/nova.conf
```

示例：

```
[DEFAULT]
my_ip=10.78.70.104
quota_instances=1000
quota_floating_ips=100
enabled_apis=osapi_compute,metadata
auth_strategy=keystone
resume_guests_state_on_host_boot=true
running_deleted_instance_timeout=300
reboot_timeout=300
instance_build_timeout=600
rescue_timeout=300
resize_confirm_window=300
ram_allocation_ratio=1.0
allow_resize_to_same_host=True
scheduler_default_filters=RetryFilter,AvailabilityZoneFilter,RamFilter,ComputeFilter,ComputeCapabilitiesFilter,ImagePropertiesFilter,ServerGroupAntiAffinityFilter,ServerGroupAffinityFilter
vif_plugging_timeout=600
firewall_driver=nova.virt.firewall.NoopFirewallDriver
allow_same_net_traffic=false
use_neutron=True
linuxnet_interface_driver=nova.network.linux_net.LinuxOVSInterfaceDriver
rpc_conn_pool_size=60
rpc_response_timeout=90
rpc_backend=rabbit
[api_database]
connection = mysql+pymysql://nova:NOVA_DBPASS@controller/nova_api
[barbican]
[cache]
[cells]
[cinder]
os_region_name=RegionOne
[conductor]
[cors]
[cors.subdomain]
```

```
[database]
connection = mysql+pymysql://nova:NOVA_DBPASS@controller/nova
[ephemeral_storage_encryption]
[glance]
api_servers=http://10.78.70.104:9292
[guestfs]
[hyperv]
[image_file_url]
[ironic]
[keymgr]
[keystone_authtoken]
auth_uri = http://10.78.70.104:5000
auth_url = http://10.78.70.104:35357
memcached_servers=controller:11211
auth_type=password
project_domain_name = default
user_domain_name = default
project_name = service
username = nova
password = ADMIN_PASS
[libvirt]
virt_type=kvm
images_type=default
images_rbd_pool=volumes
images_rbd_ceph_conf = /etc/ceph/nova-ceph.conf
rbd_user=admin
rbd_secret_uuid=81cc0d15-d09f-4526-afde-23debc9490fb
[matchmaker_redis]
[metrics]
[neutron]
url = http://10.78.70.104:9696
auth_url = http://10.78.70.104:35357
auth_type = password
project_domain_name = default
user_domain_name = default
region_name = RegionOne
project_name = service
username = neutron
password = ADMIN_PASS
service_metadata_proxy = True
metadata_proxy_shared_secret = METADATA_SECRET
timeout=90
[osapi_v21]
[oslo_concurrency]
lock_path=/var/lib/nova/tmp
[oslo_messaging_amqp]
[oslo_messaging_notifications]
driver = messagingv2
```

```
[oslo_messaging_rabbit]
rabbit_host=controller
rabbit_userid=openstack
rabbit_password=RABBIT_PASS
[oslo_middleware]
[oslo_policy]
[rdp]
[serial_console]
[spice]
[ssl]
[trusted_computing]
[upgrade_levels]
[vmware]
[vnc]
enabled=true
vncserver_listen=0.0.0.0
vncserver_proxyclient_address=$my_ip
novncproxy_base_url=http://10.78.70.104:6080/vnc_auto.html
[workarounds]
[xenserver]
```

7）初始化数据库

```
# su -s /bin/sh -c "nova-manage api_db sync" nova
# su -s /bin/sh -c "nova-manage db sync" nova
```

8）启动服务并设置开机自启动

```
# systemctl enable openstack-nova-api.service openstack-nova-cert.service openstack-nova-consoleauth.service openstack-nova-scheduler.service openstack-nova-conductor.service openstack-nova-novncproxy.service
# systemctl start openstack-nova-api.service openstack-nova-cert.service openstack-nova-consoleauth.service openstack-nova-scheduler.service openstack-nova-conductor.service openstack-nova-novncproxy.service
```

2. 在计算节点安装 Nova 相关服务

1）在各个计算节点安装 Nova 相关软件包

```
# yum install openstack-nova-compute openstack-utils numactl ceph-common
```

2）配置 nova-compute 服务使用 Ceph

```
# mkdir /etc/ceph/
```

复制控制节点/etc/ceph/下面的 ceph.*到计算节点的/etc/ceph/目录下：

```
[root@controller ceph]# ls -lh
total 20K
-rw-r--r-- 1 root root   62 Apr 20 01:31 ceph.client.admin.keyring
-rw-r--r-- 1 root root 1.4K Apr 20 01:31 ceph.conf
-rw-r--r-- 1 root root 1.2K Apr 20 01:31 nova-ceph.conf
-rwxr-xr-x 1 root root   92 Nov 21 00:17 rbdmap
```

```
-rw-r--r-- 1 root root    170 Apr 20 01:45 secret.xml
vim secret.xml
<secret ephemeral='no' private='no'>
<uuid>81cc0d15-d09f-4526-afde-23debc9490fb</uuid>
<usage type='ceph'>
<name>client.admin secret</name>
</usage>
</secret>
```

配置 secret：

 # virsh secret-define secret.xml

设置 secret valume 值，即 ceph admin 用户的 keyring 可以通过 cat ceph.client.admin.keyring 获取。

 # virsh secret-set-value --secret 81cc0d15-d09f-4526-afde-23debc9490fb --base64 AQCMdrxU+ CXeKxAAHbML+ i1XajHvXHyUq0eO9Q==

Secret value set

3）配置 libvirt，实现云主机热迁移

```
# vim /etc/libvirt/libvirtd.conf
listen_tls = 0
listen_tcp = 1
auth_tcp = "none"
# vim /etc/sysconfig/libvirtd
LIBVIRTD_ARGS="--listen"
```

4）配置计算节点 nova.conf 文件

从控制节点复制 nova.conf 文件到计算节点/etc/nova/目录下，修改：

```
[DEFAULT]
my_ip=计算节点管理ip
```

5）重启 libvirtd 和 compute 服务

 # systemctl restart libvirtd openstack-nova-compute.service

6）Nova 用户认证（计算和管理）

ssh-key nova 用户的认证还是需要的。因为在云主机冷迁移时需要两个节点之间 Nova 用户的无密码认证。

每个计算节点都要执行如下命令：

```
usermod -s /bin/bash nova
su nova
mkdir -p /var/lib/nova/.ssh
cd /var/lib/nova/
cat > .ssh/config <<EOF
Host *
StrictHostKeyChecking no
```

```
UserKnownHostsFile=/dev/null
EOF
cd .ssh/
ssh-keygen -f id_rsa -b 1024 -P ""
cp id_rsa.pub authorized_keys
```

最后将所有节点的 authorized_keys 整理为一个，存放到每个节点的/var/lib/nova/.ssh/下。

13.4.8 在控制节点安装 Neutron 相关服务

1. 在数据库中创建 neutron 数据库并配置 neutron 数据库的访问权限

```
# mysql -u root -p
```

配置数据库：

```
CREATE DATABASE neutron;
GRANT ALL PRIVILEGES ON neutron.* TO 'neutron'@'localhost' IDENTIFIED BY 'NEUTRON_DBPASS';
GRANT ALL PRIVILEGES ON neutron.* TO 'neutron'@'%' IDENTIFIED BY 'NEUTRON_DBPASS';
GRANT ALL PRIVILEGES ON neutron.* TO 'neutron'@'controller' IDENTIFIED BY 'NEUTRON_DBPASS';
```

2. 在 Keystone 中配置 Neutron 用户、租户、角色、端点等信息

```
# source admin-openrc
# openstack user create --domain default --password NEUTRON_PASS neutron
# openstack role add --project service --user neutron admin
# openstack service create --name neutron \
    --description "OpenStack Networking" network
# openstack endpoint create --region RegionOne \
    network public http://10.78.70.104:9696
# openstack endpoint create --region RegionOne \
    network internal http://10.78.70.104:9696
# openstack endpoint create --region RegionOne \
    network admin http://10.78.70.104:9696
```

3. 在控制节点安装网络组建服务相关软件包

```
#yum install openstack-neutron openstack-neutron-ml2 which openvswitch openstack-neutron-openvswitchipset
```

4. 配置 Neutron 相关配置文件

```
# vim /etc/neutron/neutron.conf
```

示例：

```
[DEFAULT]
state_path = /var/lib/neutron
auth_strategy = keystone
core_plugin = ml2
```

```
service_plugins = router
allow_overlapping_ips = True
notify_nova_on_port_status_changes = true
notify_nova_on_port_data_changes = true
api_workers = 20
rpc_workers = 20
dhcp_agents_per_network = 2
router_scheduler_driver = neutron.scheduler.l3_agent_scheduler.LeastRoutersScheduler
allow_automatic_l3agent_failover = True
rpc_backend = rabbit
[agent]
availability_zone = nova
[cors]
[cors.subdomain]
[database]
connection = mysql+pymysql://neutron:NEUTRON_DBPASS@controller/neutron
max_pool_size = 100
max_overflow = 30
[keystone_authtoken]
auth_uri = http://10.78.70.104:5000
auth_url = http://10.78.70.104:35357
memcached_servers = controller:11211
auth_type = password
project_domain_name = default
user_domain_name = default
project_name = service
username = neutron
password = ADMIN_PASS
[matchmaker_redis]
[nova]
auth_url = http://10.78.70.104:35357
auth_type = password
project_domain_name = default
user_domain_name = default
region_name = RegionOne
project_name = service
username = nova
password = ADMIN_PASS
[oslo_concurrency]
lock_path = $state_path/lock
[oslo_messaging_amqp]
[oslo_messaging_notifications]
driver = messagingv2
[oslo_messaging_rabbit]
rabbit_host = controller
rabbit_userid = openstack
rabbit_password = RABBIT_PASS
[oslo_policy]
```

[quotas]
[ssl]

配置 Modular Layer plug-in：

vim /etc/neutron/plugins/ml2/ml2_conf.ini

示例：

```
[DEFAULT]
[ml2]
type_drivers = flat,vxlan
tenant_network_types = vxlan,flat
mechanism_drivers = openvswitch,l2population
extension_drivers = port_security
[ml2_type_flat]
flat_networks = flat-network
[ml2_type_geneve]
[ml2_type_gre]
[ml2_type_vlan]
[ml2_type_vxlan]
vni_ranges = 1:100000
vxlan_group = 239.1.1.1
[securitygroup]
firewall_driver = neutron.agent.linux.iptables_firewall.OVSHybridIptablesFirewallDriver
enable_security_group = true
enable_ipset = true
[ovs]
enable_tunneling = True
tunnel_type = vxlan
bridge_mappings = flat-network:br-flat
external_network_bridge = br-flat
local_ip = 10.0.0.101
[agent]
l2_population = True
tunnel_types = vxlan
```

配置软连接：

 # ln -s /etc/neutron/plugins/ml2/ml2_conf.ini /etc/neutron/plugin.ini

5. 初始化数据库

**# su -s /bin/sh -c "neutron-db-manage --config-file /etc/neutron/neutron.conf \
　--config-file /etc/neutron/plugins/ml2/ml2_conf.ini upgrade head" neutron**

6. 配置 l3-agent、metadata-agent、dhcp-agent 服务

1）配置 l3-agent

 # vim /etc/neutron/l3_agent.ini
```
[DEFAULT]
interface_driver = neutron.agent.linux.interface.OVSInterfaceDriver
```

```
external_network_bridge =
agent_mode = dvr_snat
[AGENT]
```

2）配置 dhcp 组件

```
# vim /etc/neutron/dhcp_agent.ini
[DEFAULT]
interface_driver = neutron.agent.linux.interface.OVSInterfaceDriver
dhcp_driver = neutron.agent.linux.dhcp.Dnsmasq
enable_isolated_metadata = True
[AGENT]
```

3）配置元数据组建

```
# vim /etc/neutron/metadata_agent.ini
[DEFAULT]
nova_metadata_ip = controller
metadata_proxy_shared_secret = METADATA_SECRET
[AGENT]
```

权限说明：

/etc/neutron/的权限都是 root:neutron
chown -R root:neutron /etc/neutron/*

7. 开启 OVS 服务

```
# systemctl enable openvswitch.service
# systemctl start openvswitch.service
```

8. 配置桥接

```
# ovs-vsctl show
# ovs-vsctl add-br br-int
# ovs-vsctl add-br br-flat
# ip link list
# ovs-vsctl add-port br-flat eth2
# ethtool -K eth2 gro off
# ethtool -K eth3 gro off
```

9. 网卡配置

1）管理网卡

```
# vim ifcfg-eth0
BOOTPROTO=static
DEVICE=eth0
ONBOOT=yes
TYPE=Ethernet
IPADDR=10.78.70.104
NETMASK=255.255.0.0
GATEWAY=10.78.70.1
```

2）存储网卡

```
# vim ifcfg-eth1
BOOTPROTO=static
DEVICE=eth1
ONBOOT=yes
TYPE=Ethernet
IPADDR=10.10.244.101
NETMASK=255.255.255.0
```

3）Flat 网卡

```
# vim ifcfg-eth2
DEVICE=eth2
TYPE=OVSPort
DEVICETYPE=ovs
BOOTPROTO=none
ONBOOT=yes
OVS_BRIDGE=br-flat
IPV6INIT=no
vim ifcfg-br-flat
DEVICE=br-flat
TYPE=OVSBridge
DEVICETYPE=ovs
ONBOOT=yes
BOOTPROTO=none
```

4）配置 VxLAN 网卡

```
# vim ifcfg-eth3
BOOTPROTO=static
DEVICE=eth3
ONBOOT=yes
TYPE=Ethernet
IPADDR=10.0.0.101
NETMASK=255.255.0.0
```

10. 重启网络相关服务

```
# /etc/init.d/network restart
```

11. 配置好 neutron-server 和 openvswitch-agent 都使用 plugin.ini

```
# sed -i 's,plugins/openvswitch/ovs_neutron_plugin.ini,plugin.ini,g' /usr/lib/systemd/system/neutron-server.service
# sed -i 's,plugins/ml2/openvswitch_agent.ini,plugin.ini,g' /usr/lib/systemd/system/neutron-openvswitch-agent.service
# systemctl daemon-reload
```

启动服务并设置开机自启动。
开启服务和配置开机自启动。

```
# systemctl enable neutron-server
# systemctl start neutron-server
# systemctl enable neutron-openvswitch-agent.service neutron-l3-agent.service neutron-dhcp-agent.service neutron-metadata-agent.service neutron-openvswitch-agent.service
# systemctl restart neutron-openvswitch-agent.service neutron-l3-agent.service neutron-dhcp-agent.service neutron-metadata-agent.service neutron-openvswitch-agent.service
```

验证 source admin-openrc：

```
# neutron agent-list
```

13.4.9 在计算节点安装 Neutron 相关服务

1. 在计算节点安装网络组建相关软件包

```
# yum install openstack-neutron openvswitch openstack-neutron-openvswitchipset
```

2. 配置 Neutron 相关配置文件

```
# vim /etc/neutron/neutron.conf
```

示例：

```
[DEFAULT]
state_path = /var/lib/neutron
auth_strategy = keystone
core_plugin = ml2
service_plugins = router
allow_overlapping_ips = True
notify_nova_on_port_status_changes = true
notify_nova_on_port_data_changes = true
api_workers = 20
rpc_workers = 20
dhcp_agents_per_network = 2
router_scheduler_driver = neutron.scheduler.l3_agent_scheduler.LeastRoutersScheduler
allow_automatic_l3agent_failover = True
rpc_backend = rabbit
[agent]
availability_zone = nova
[cors]
[cors.subdomain]
[database]
connection = mysql+pymysql://neutron:NEUTRON_DBPASS@controller/neutron
max_pool_size = 100
max_overflow = 30
[keystone_authtoken]
auth_uri = http://10.78.70.104:5000
auth_url = http://10.78.70.104:35357
memcached_servers = controller:11211
auth_type = password
```

```
project_domain_name = default
user_domain_name = default
project_name = service
username = neutron
password = ADMIN_PASS
[matchmaker_redis]
[nova]
auth_url = http://10.78.70.104:35357
auth_type = password
project_domain_name = default
user_domain_name = default
region_name = RegionOne
project_name = service
username = nova
password = ADMIN_PASS
[oslo_concurrency]
lock_path = $state_path/lock
[oslo_messaging_amqp]
[oslo_messaging_notifications]
driver = messagingv2
[oslo_messaging_rabbit]
rabbit_host = controller
rabbit_userid = openstack
rabbit_password = RABBIT_PASS
[oslo_policy]
[quotas]
[ssl]
```

3. 配置 Modular Layer plug-in

 # vim /etc/neutron/plugins/ml2/ml2_conf.ini

示例：

```
[DEFAULT]
[ml2]
type_drivers = flat,vxlan
tenant_network_types = vxlan,flat
mechanism_drivers = openvswitch,l2population
extension_drivers = port_security
[ml2_type_flat]
flat_networks = flat-network
[ml2_type_geneve]
[ml2_type_gre]
[ml2_type_vlan]
[ml2_type_vxlan]
vni_ranges = 1:100000
vxlan_group = 239.1.1.1
[securitygroup]
```

```
firewall_driver = neutron.agent.linux.iptables_firewall.OVSHybridIptablesFirewallDriver
enable_security_group = true
enable_ipset = true
[ovs]
enable_tunneling = True
tunnel_type = vxlan
bridge_mappings = flat-network:br-flat
external_network_bridge = br-flat
local_ip = 10.0.0.102
[agent]
l2_population = True
tunnel_types = vxlan
```

配置连接：

ln -s /etc/neutron/plugins/ml2/ml2_conf.ini /etc/neutron/plugin.ini

初始化数据库权限说明：

/etc/neutron/的权限都是 root:neutron
chown -R root:neutron /etc/neutron/*

4. 开启 OVS 服务

systemctl enable openvswitch.service
systemctl start openvswitch.service

5. 配置桥接

ovs-vsctl show
ovs-vsctl add-br br-int
ovs-vsctl add-br br-flat
ip link list
ovs-vsctl add-port br-flat eth2
ethtool -K eth2 gro off
ethtool -K eth3 gro off

6. 网卡配置

1）管理网卡

```
# vim ifcfg-eth0
BOOTPROTO=static
DEVICE=eth0
ONBOOT=yes
TYPE=Ethernet
```

2）IPADDR=计算节点管理 IP

NETMASK=255.255.0.0
GATEWAY=10.78.70.1

3）存储网卡

vim ifcfg-eth1

```
BOOTPROTO=static
DEVICE=eth1
ONBOOT=yes
TYPE=Ethernet
IPADDR=10.10.244.101
NETMASK=255.255.255.0
```

4）Flat 网卡

```
# vim ifcfg-eth2
DEVICE=eth2
TYPE=OVSPort
DEVICETYPE=ovs
BOOTPROTO=none
ONBOOT=yes
OVS_BRIDGE=br-flat
IPV6INIT=no
vim ifcfg-br-flat
DEVICE=br-flat
TYPE=OVSBridge
DEVICETYPE=ovs
ONBOOT=yes
BOOTPROTO=none
```

5）配置 VxLAN 网卡

```
# vim ifcfg-eth3
BOOTPROTO=static
DEVICE=eth3
ONBOOT=yes
TYPE=Ethernet
IPADDR=10.0.0.102
NETMASK=255.255.0.0
```

7. 重启网络相关服务

`# /etc/init.d/network restart`

配置好 neutron-server 和 openvswitch-agent 都使用 plugin.ini：

```
# sed -i 's,plugins/openvswitch/ovs_neutron_plugin.ini,plugin.ini,g' /usr/lib/systemd/system/neutron-server.service
# sed -i 's,plugins/ml2/openvswitch_agent.ini,plugin.ini,g' /usr/lib/systemd/system/neutron-openvswitch-agent.service
# systemctl daemon-reload
```

生效服务并启动服务，重启服务并设置服务开机自启动：

```
# systemctl enable neutron-server
# systemctl start neutron-server
# systemctl enable neutron-openvswitch-agent.service
# systemctl restart neutron-openvswitch-agent.service
```

13.4.10　安装 Dashboard

1. 在控制节点安装 Dashboard

`#yum install openstack-dashboard`

2. 配置 Dashbaord

`# vim /etc/openstack-dashboard/local_settings`

（整个文件内容比较多，所以只记录修改的部分。）

```
OPENSTACK_HOST = "controller"
OPENSTACK_KEYSTONE_URL = "http://%s:5000/v3" % OPENSTACK_HOST
OPENSTACK_KEYSTONE_MULTIDOMAIN_SUPPORT = True
ALLOWED_HOSTS = ['*',]
SESSION_ENGINE = 'django.contrib.sessions.backends.cache'
CACHES = {
    'default': {
            'BACKEND': 'django.core.cache.backends.memcached.MemcachedCache',
            'LOCATION': 'controller:11211',
    }
}
OPENSTACK_API_VERSIONS = {
    #"data-processing": 1.1,
"identity": 3,
"image": 2,
"volume": 2,
"compute": 2,
}
OPENSTACK_KEYSTONE_DEFAULT_DOMAIN = "default"
OPENSTACK_KEYSTONE_DEFAULT_ROLE = "user"
OPENSTACK_NEUTRON_NETWORK = {
    'enable_router': True,
    'enable_quotas': True,
    'enable_ipv6': True,
    'enable_distributed_router': False,
    'enable_ha_router': False,
    'enable_lb': True,
    'enable_firewall': True,
    'enable_vpn': True,
'enable_fip_topology_check': True,
…
}
TIME_ZONE = "Asia/Shanghai"
```

3. 启动服务并设置开机自启动

`# systemctl enable httpd.service memcached.service`
`# systemctl start httpd.service memcached.service`

13.5　OpenStack 运维案例

作为一名运维工程师，其主要工作是负责维护并确保整个服务的高可用性，并不断优化系统架构、提升部署效率、优化资源利用率以提高整体的 ROI。假如你是一名 OpenStack 运维工程师，应该从哪些方面去保证所搭建的平台能平稳、可靠地被使用？这里，我们从服务状态方面入手，做一个自动化运维脚本。该脚本主要实现定时对 OpenStack 相关服务状态进行轮询，对 down 掉的服务能够重启，同时能够发送邮件提醒，在邮件的内容上能够说明在什么时间重启了什么服务，重启后，该服务是否正常。这里服务重启的时间戳就显得很有必要，运维人员可以根据时间戳来查看该服务 down 掉的日志，通过日志可以分析出该服务 down 掉的原因。基于上面的思想，需要做以下几个方面的准备。

1. 获取 OpenStack 相关服务

1）控制节点服务

```
[root@controller ~]#
openstack-service list
neutron-clb-agent
neutron-dhcp-agent
neutron-l3-agent
neutron-metadata-agent
neutron-metering-agent
neutron-openvswitch-agent
neutron-qos-agent
neutron-server
neutron-vpn-agent
openstack-cinder-api
openstack-cinder-scheduler
openstack-cinder-volume
openstack-glance-api
openstack-glance-registry
openstack-nova-api
openstack-nova-cert
openstack-nova-conductor
openstack-nova-consoleauth
openstack-nova-novncproxy
openstack-nova-scheduler
```

另外加上数据库服务和消息队列服务。

2）计算节点服务

计算节点相对来说服务较少，重要的有以下三个：

```
neutron-openvswitch-agent
openstack-nova-compute
libvirtd
```

邮件发送。

邮件发送配置：

```
hii=/usr/local/src/hii
logs=/usr/local/src/hii/logs
cd $logs
wget http://caspian.dotconf.net/menu/Software/SendEmail/sendEmail-v1.56.tar.gz
tar zxvf sendEmail-v1.56.tar.gz
cp $logs/sendEmail-v1.56/sendEmail /usr/sbin/
chmod +x /usr/sbin/sendEmail
cat $hii/openrc.sh
sendEmail="/usr/sbin/sendEmail"
sender=send@chinac.com
recipient=recive@chinac.com
smtp=smtp.chinac.com
cipher=xxx
```

编写测试脚本：

cat /opt/service_check.sh
```
#!/bin/bash
hii=/usr/local/src/hii
logs=/usr/local/src/hii/logs
source $hii/openrc.sh
for services in vsftpd tuned
do
service_status=`systemctl status ${services}.service |grep Active |awk '{print $2}'`
if[ $service_status !='active']
then
systemctl status $services > ${logs}/${services}.txt
mail -s '$services is not running' test@qq.com < $logs/${services}.txt
fi
done
```

2．完成运维脚本

1）编写脚本

service_check.sh
```
#!/bin/bash
HOST_IP=10.78.70.104
date=`date +%Y%m%d%H%M%S`
mkdir -p /usr/local/src/hii/logs
hii=/usr/local/src/hii
logs=/usr/local/src/hii/logs
sendEmail=/usr/sbin/sendEmail
for service in `cat /root/service_check/*`
   do
      service_status=`systemctl status ${service} |grep Active |awk '{print $2}'`
      if [ $service_status != 'active' ]
         then
            date > ${logs}/$date-${service}.txt
```

```
            systemctl status $service >> ${logs}/$date-${service}.txt

            systemctl restart $service
            sleep 10
            systemctl status $service >> ${logs}/$date-${service}.txt
            if [ `systemctl status ${service} |grep Active |awk '{print $2}'` != 'active' ]
              then
                 object='failed'
            else
                 object='started after restart'
            fi
            $sendEmail -u "$HOST_IP controller $service $object" -t  -f  -o message-content-type=text -o message-charset=utf8 -o tls=no -o message-file=$hii/logs/$date-${service}.txt -s   -xu     -xp
              fi
        done
```

将计算节点和控制节点要监控的服务名字放到/root/service_check/下的某个文件即可。

2）设置定时任务

[root@controller ~]# crontab -e
添加*/5 * * * * /root/hii/service_check.sh

每隔5min会执行一次服务检测的脚本。

课后习题参考答案

第 1 章

1. 电厂模式、效用计算、网格计算、云计算
2. 快速满足业务需求低成本；绿色节能提高资源管理效率
3. 应用层、平台层、基础设施层、虚拟化层
4. IaaS：是 Infrastructure as a Service，基础设施即服务。
SaaS：是 Software as a Service，软件运营服务模式，简称为软营模式，软件即服务。
PaaS：是 Platform as a Service，平台即服务。
5. Keystone、Heat、Horizon、Neutron、Cinder、Nova、Ceilometer、Glance、Swift

第 2 章

1. 全虚拟化、半虚拟化；
服务器虚拟化、桌面虚拟化、应用虚拟化
2. 开源全虚拟化；
CPU 虚拟化指令集
3. 用户网络、虚拟网桥
4. virshdominfo KVM123
5. lsmod |grep kvm

第 3 章

1. B
2. C
3. 认证、权限管理
4. 监控、计量
5. Swift
6. Keystone、Glance、Horizon 以及 Nova 和 Neutron 中管理相关的组件；
SQL 数据库、消息队列和网络时间服务

7. Nova、Neutron、Nlance、Cinder、Keystone、Horizon。比如 Nova 的作用是管理 VM 的生命周期；Ceilometer 的作用是为 OpenStack 提供监控和计量服务。

第 4 章

1．身份认证；

身份认证；

域、用户、凭证、认证、令牌、租户、服务、角色、Keystone 客户端、策略、端点

2．su -s /bin/sh -c "keystone-manage db_sync" keystone

3．openssl rand -hex 10

4．35357

5．openstack project create --domain default --description "cqcetProject" cqcet

6．CREATE DATABASE cqcet;

GRANT ALL PRIVILEGES ON cqcet.* TO 'cqcet'@'localhost' IDENTIFIED BY '2015170777';

GRANT ALL PRIVILEGES ON cqcet.* **TO 'cqcet'**@'%' IDENTIFIED BY '2015170777';

第 5 章

1．Glance

2．openstack user create --domain default --password-prompt glance

3．9292

4．su -s /bin/sh -c "glance-manage db_sync" glance

5．openstack image create "CIRROS" --file cirros-0.3.4-x86_64-disk.img --disk-format qcow2 --container-format bare --public

第 6 章

1．API Server、Message Queue、Compute Worker、Network Controller、Volume Workers、Scheduler

2．数据库、队列

3．su -s /bin/sh -c "nova-manage api_db sync" nova

4．egrep -c '(vmx|svm)' /proc/cpuinfo

5．Systemctl start libvirtd.service openstack-nova-compute.service

第 7 章

1．su -s /bin/sh -c "neutron-db-manage --config-file /etc/neutron/neutron.conf --config-file /etc/neutron/plugins/ml2/ml2_conf.ini upgrade head" neutron

2．http://controller:9696

3．ln -s /etc/neutron/plugins/ml2/ml2_conf.ini /etc/neutron/plugin.ini

4．neutron agent-list

第 8 章

1．B

2．D

3．A

4．对象存储

5．XFS

6．swift stat

7．swift-ring-builder container.builder

第 9 章

1．Web、OpenStack

2．计算、存储、网络资源

第 10 章

1．A

2．C

3．A

4．su -s /bin/sh -c "cinder-manage db sync" cinder

5．cinder service-list

6．cinder list

7．openstack endpoint create --region RegionOne volume public http://controller: 8776/v1/%\(tenant_id\)s

openstack endpoint create --region RegionOne volumev2 public http://controller: 8776/v2/%\(tenant_id\)s

第 11 章

1．C

2．A

3．systemctl start openstack-heat-api.service openstack-heat-api-cfn.service openstack- heat-engine.service

4．openstack orchestration service list

第 12 章

1．略

2．见第 12 章 12.2 节的内容

反侵权盗版声明

电子工业出版社依法对本作品享有专有出版权。任何未经权利人书面许可，复制、销售或通过信息网络传播本作品的行为，歪曲、篡改、剽窃本作品的行为，均违反《中华人民共和国著作权法》，其行为人应承担相应的民事责任和行政责任，构成犯罪的，将被依法追究刑事责任。

为了维护市场秩序，保护权利人的合法权益，我社将依法查处和打击侵权盗版的单位和个人。欢迎社会各界人士积极举报侵权盗版行为，本社将奖励举报有功人员，并保证举报人的信息不被泄露。

举报电话：（010）88254396；（010）88258888
传　　真：（010）88254397
E-mail：　dbqq@phei.com.cn
通信地址：北京市海淀区万寿路173信箱
　　　　　电子工业出版社总编办公室
邮　　编：100036